Engineers *becoming* Managers

from the Classroom to the Boardroom

By Peter C. Hughes
BASc, PEng, PhD, MBA

Emeritus Professor, Faculty of Applied Science and Engineering
Emeritus Professor, Joseph L. Rotman School of Management
University of Toronto

Copyright © 2006 by Peter C. Hughes.
Library of Congress Number: 2006908846
ISBN 10: Hardcover 1-4257-3650-5
Softcover 1-4257-3649-1

ISBN 13: Hardcover 978-1-4257-3650-7
Softcover 978-1-4257-3649-1

All rights reserved. No part of this book may be reproduced or transmitted in any form or by any means, electronic or mechanical, including photocopying, recording, or by any information storage and retrieval system, without permission in writing from the copyright owner.

This book was printed in the United States of America.

To order additional copies of this book, contact:
Xlibris Corporation
1-888-795-4274
www.Xlibris.com
Orders@Xlibris.com

Contents

Preface v

Chapter 1: This Book's Motivation and Goals 1
    1.1   One Size Does Not Fit All 1
    1.2   The Engineering-Management Connection 2
           Engineers Whose Careers Are Dynamic; What Engineers Actually Do; Management Tends To Come With Age; Skills Engineers Believe They Need
    1.3   The Three Pillars Of The Engineering Profession 9
    1.4   This Book Will Help Three Groups Of Readers 11
           Scenario 1: The Reader Is In Engineering School; Scenario 2: The Reader Has Just Started An Engineering Career; Scenario 3: The Reader Is An Experienced Engineering Manager
    1.5   Summary Of Book Structure 16

PART I: FOR ENGINEERING STUDENTS

Chapter 2: Applied-Science Bias in Engineering School 19
    2.1   Further Description Of The Bias 20
           Status Among Possible Careers; One Pillar Is Weak And Neglected
    2.2   Reasons For The Bias 23
           Reason 1: The Urge To Clone Oneself; Reason 2: It's Hard To Lose Your Best; Reason 3: Extrinsic Influences; Reason 4: What Gets Rewarded Gets Done; Reason 5: Papers, Sabbaticals, Etc.; Reason 6: Prior Selection
    2.3   General Benefits Of Undergraduate Engineering 28
           Benefit 1: New (Post-secondary) Challenges Are Salutary; Benefit 2: Math Classes Offer Special Benefits; Benefit 3: Science Classes (Should) Create Healthy Skepticism
    2.4   New Curriculum Options Should Be Offered 30
           Inbreeding Can't Be Good; Letting The Light Of The Profession Flood The Cloister; Engineering Professors As Engineering Consultants
    2.5   New Program Options Should Be Offered 35
           The Applied Science Stream; The Design Stream; The Management Stream; The MEng Stream; Time On The Job

Chapter 3: Important Decisions in Engineering School 41

3.1 Undergraduate Career Choices 42
Program Segmentation—The Ideal Curriculum Structure; Typical Actual Undergraduate Curriculum

3.2 Graduate Career Choices 47
MBA vs. PhD; Immediate Employment vs. PhD

3.3 It's Not Rocket Science—it's More Difficult! 53
Mathematics vis-à-vis Management; Which Is The Bird Course?; Side Benefits To Studying Management

3.4 A Constellation Of Professions 58
Principle Of Vocophilia; Other Key Professions

3.5 Technology As Viewed By Other Professions 62
Engineers And Technologists; Engineering Professors; Accountants; Lawyers; Business Valuators And Investors

3.6 In Search Of A Business Model 68
In Praise Of Block Diagrams; Is There A Generic Business Model?; Basic Financial Business Model; A Call To Arms

Chapter 4: The Long Path 73

4.1 Abstract Mathematics 76
Example 1: The Number π; Example 2: Fermat's Last Theorem; Example 3: Guthrie's Four-Color Conjecture; Example 4: Riemann's Hypothesis; Conjecture On Utility Of Current Abstract Mathematics Work

4.2 Applied Mathematics 87
Applied Mathematics; Venn Diagram

4.3 Science 89
Scientific Method Example 1: Water That Remembers; Scientific Method Example 2: Energy From The Kitchen Sink; Scientific Method Example 3: The Cause Of Stomach Ulcers; Unsolved Problems In Science, And Their Application

4.4 Applied Science And Engineering 97
Example 1: Structural Statics; Example 2: Structural Dynamics; Approaches To Approximation; Engineering Analysts At Their Best—the Finite Element Method

4.5 Next Steps In The Long Path 104

Chapter 5: Marketing Concepts 107

5.1 Basic Marketing Concepts 108
Technology Push vs. Market Pull; Segmentation; Positioning; Differentiation; Niche Marketing

5.2 More About Products 115
What Is A Product?; Price; Value; Quality; Green Belts, Black Belts, And The Hidden Factory

5.3 Lifetime, Channels, And Newness 124
Product Lifetime; Channels Of Distribution; How New Is "New"?

5.4 Marketing Of High-technology Products 128
   The Many Faces Of Technology; Product Technology vs. Process Technology; Moore's Segmentation; The Chasm

Chapter 6: Politics at Work 139

6.1 Corporate Politics: An Inescapable Reality 140
   Politics Is Competitive By Nature; A Turn-on? Or A Turn-off?; Purpose Of This Chapter

6.2 Understanding Politics 143
   Politics vis-à-vis Policies; The Good, The Bad And The Ugly; Politics Reframed; The Best Strategy

6.3 Six Common Tactics For Winning At Corporate Politics 147
   Tactic 1—External Attribution Of Blame; Tactic 2—Controlling Information; Tactic 3—Credit Appropriation; Tactic 4—Building Networks And Coalitions; Tactic 5—Crafting Obligations; Tactic 6—Spinning Images

6.4 Finding Your Political Fit 158
   Degree Of Politicization; Which Degree Of Politicization Is Preferable?

6.5 In Closing—two Thoughts 161

PART II: FOR YOUNG PROFESSIONAL ENGINEERS

Chapter 7: Some Risks are Worth Taking 165

7.1 What Is Risk? 166
   Risk Requires A Probabilistic Approach; An Example Of Mathematical Probability

7.2 Human Responses To Risk 171
   The First Human Response To Risk: Aversion; Aversion To Change; The Second Human Response To Risk: Underestimation; Simple Mathematical Insights

7.3 First Component Of Business Risk: Technology Risk 179
   Technology Risk; Technology Readiness; Examples Of The Technology Readiness Levels

7.4 Second Component Of Business Risk: Market Risk 183
   Market Risk; Market Readiness; Technology Risk And Market Risk Combined; Technology Push vs. Market Pull

7.5 Third Component Of Business Risk: Management Risk 189
   Combining Management (And Other) Risks

7.6 The Risk-reward Relationship 191
   Risk-reward Principle (Strong Form); Risk-reward Examples; SWOTing Risks

v

Chapter 8: Accountancy 199

8.1 Engineers vis-à-vis Accountants 199
Engineers Don't Naturally Understand The Accounting Culture; Subject With No Theory; Some Strengths And Limitations

8.2 An Extended Example Of Financial Accounting 203

8.3 Financial And Managerial Accounting 210
Financial Reporting Is Done For Outsiders; Managerial Accounting Is Done For Insiders

8.4 The Income Statement 212
EBITDA—Just A Scam?

8.5 The Balance Sheet 220
Risk And The Balance Sheet

8.6 Cash Flow Statement; And Notes 227
Utility Of The Cash Flow Statement; Notes To The Financial Statements; Revealing Ratios

8.7 Managerial Accounting 231
Breakeven; Activity-based Costing

Chapter 9: Innovation 235

9.1 What Is Innovation? 236
Role Of Innovation In The Economy; Not Everyone Benefits Immediately; Much More Than Invention; Innovation Is Everyone's Business

9.2 Innovation Is Long, Hard Work 243
The Timeframe For Innovation Can Be Decades; Peter Drucker On Innovation; Example—The Six-Hat Process

9.3 Research And Development 248
The Role Of Universities In Fostering (Technology) Innovation; Technology Transfer From Universities; Pre-competitive Research; Public Sector R&D; Measuring R&D Impact

9.4 Patterns Of Innovation 258
The S-Curve; Generations Of Technology; The Learning Curve; Process Innovation Vis-à-Vis Product Innovation

9.5 Innovation Frameworks 262
Incubators For Innovation; Technology Clusters

Chapter 10: Intellectual Capital 269

10.1 What Is The Value Of A Corporation?
—Accountancy's Current Answer 269
Tangible Assets ; That Intangible Word, "Intangible"; Accountancy's Treatment Of Intangible Assets; Intangible Assets—A Valuation Challenge That Must Be Met; A Counter-Argument

10.2 What Is The Value Of A Corporation?—A More Accurate Answer 275
What Does Mr Market Say?; Intellectual Capital; Other Aspects Of Value

10.3 Sources And Components Of Intellectual Capital 283
Operating System For A Network Of Brains; Best Brains; The Skandia Model; Intellectual Property (IP); Other Intellectual Capital; Accountants Are Taking Note

10.4 Knowledge Management..292
The Refinement Of Knowledge; The Business As Control System; Balanced Scorecard

10.5 Attracting And Retaining Quality Personnel 301
Compensation; Benefits; Short-term Incentives; Stock Options

PART III: ENGINEERS WHO BECOME LEADERS AND SENIOR MANAGERS

Chapter 11: Leadership 311

11.1 Some Engineers Make Outstanding Managers..312
A Helpful Analogy; Anti-management Bias In The Engineering-school Culture; Playing The Management Card In Mid-career; Anti-management Bias In Typical Engineering Associations; Calling A Duck A Duck

11.2 Business Leadership 321
Leadership Is More Than Management; Leadership Defined; Leadership And Management—Good And Bad; Intellectual Capital Revisited; Special Problems (And Benefits) With Engineers Becoming Managers

11.3 The Engineer → Management Transition: A Checklist 329
A Personal Experience; The Six Key Characteristics Of Effective Leadership; Leadership Characteristic 1: Verbal (Oral + Written) Communication; Leadership Characteristic 2: Leading People; Leadership Characteristic 3: Being Results Driven (Plus); Leadership Characteristic 4: Business Acumen; Leadership Characteristic 5: Political Astuteness; Leadership Characteristic 6: Leading Change

11.4 A Final Word 339

Chapter 12: Entrepreneurship 341
Startups Under Discussion; Connections With Earlier Chapters; Entrepreneurship vis-à-vis Innovation

12.1 Size Does Matter...346
Companies Have Phases Of Growth; Examples: Communications And Financial Resources; A Three-stage Growth Model; A Six-stage Growth Model; Many Opportunities To Foul It Up

12.2 The Essential Species: Entrepreneur..352
How Do Entrepreneurs Think?; Can Engineering Students Be Entrepreneurs?; Three Common Entrepreneurial Weaknesses; Can The Entrepreneur Scale?; Can The Business Scale?

12.3 Acquiring Much-needed Cash 366
Stages Of Investment; Seed Investment; Angel Investment; Venture Capital Investment; Further Stages Of Investment; General Stance On Outside Investment

Chapter 13: Governance **375**

13.1 Engineers In Senior Management 377
Senior Rungs On The Management Ladder; Officers Of The Corporation; Chief Operating Officer (COO)

13.2 Serving As A Director In Governance 383
Engineers On The BoD; The Concept Of Governance; Fiduciary Behavior; Legal Liability Of Individual Directors; High-risk Behavior For Directors; Due Diligence; Business Judgment Decisions

13.3 The Governance Team 393
Critical Bod Roles; BoD Makeup; Women On The BoD; Subverting Outside Directors; Chairman & CEO: One Role Or Two?; BoD Committees; CEO Evaluation And BoD Evaluation; Shareholders vis-à-vis Stakeholders; Further Remarks On Governance

13.4 The Strategic Imperative 409
The Endless Quest For A Good Strategy; Vision, Mission, Purpose . . . ; SWOT (Again); BoD Role In Corporate Strategy; Technology Roadmapping

Dedication

The author is mindful that, had not the following
two highly competent physicians intervened in the
progress of life-threatening diseases,
he would not have been available to write this book:

Felix L. W. Liu, M.D.
respirologist and internist (for pneumonia, 2000)

Joel G. DeKoven, M.D.
dermatologist (for squamous-cell carcinoma, 2002)

Preface

This book rests on a Tale of Three Cultures—these cultures being: applied science, engineering, and management. While these plainly overlap to a degree, a person cannot move from success in one to success in another without considerable effort, dedication and talent. Clearly, an understanding of these cultural differences is essential to engineers whose career goal is to evolve into top-level managers. The first step in gaining such understanding is to admit that these three cultures are quite distinct. The applied science culture is typified by the engineering school; the engineering culture is typified by the company engineering design office; and the management culture is typified by the senior management team and the boardroom.

The Three Cultures

The older one gets, the more one realizes the enormous importance of "culture" to almost every important human issue, and the topic of engineers becoming managers is certainly no exception. The *culture* of a group is the *set of all common traits, responses, values, beliefs, priorities, attitudes and behaviors* which characterize that group. There will inevitably be some degree of variation among individuals in the group, but significant deviations by group members will lead to dissonances; if a putative member of a culture exhibits cultural traits that are not normative, such a member will be said to be "not really one of us, more like one of them."

A group's culture is plainly of the utmost importance to that group, but a culture's most interesting characteristic is that it is not normally passed on, from older group members to younger ones, by some public, written declaration of principles, policies, or intentions, or some other type of printed manifesto. It is transmitted instead by a thousand subtle messages, most being nonverbal.

While these general comments on culture apply to all human activity, in this book I wish to stimulate new thoughts on these issues in a specific instance, namely, careers that are touched by the three cultures noted: applied science, engineering, and management. I fully recognize that a great many books have been written within each of these three cultures individually—although, because these books are centered on only one culture, the "culture" language is never used (an example of the subtlety of culture). My hope for this book is that some helpful light can be shed on all three "cultural imperatives" at once—or, more precisely, on the career considerations for someone who must navigate through all three cultures in their life's career journey.

We can safely assume that the three cultures I have highlighted are encountered in the order mentioned, namely, applied science (Culture AS), then (real) engineering (Culture E), and finally management (Culture M). Although students, even in secondary school, are brought into some meaningful contact with mathematics and science—typical of Culture AS—the *gestalts* associated with Cultures E and M have not yet begun.

For any engineer in a normal engineering environment, some management is simply part of the job; if an engineer attempts to eschew all forms of management, she is likely to be derelict in her duty and will certainly limit her professional opportunities.

Preface

This professional fact ineluctably links two of the professions, and thus two of the cultures. In fact, I shall argue herein that "management" is not just some obscure nuisance with which engineers must somehow grapple while seeking to minimize it. On the contrary, the profession (and therefore the culture) of management is a critical component of the engineering profession and culture. Chapter 1 demonstrates that engineers (Culture E) spend a good deal of their time involved with some form of management (Culture M). Though Cultures E and M are different in many important respects, the former does not have the luxury of pretending that the latter doesn't exist.

Part I Of This Book
Having thus briefly established in Chapter 1 the inseparability of engineering (Culture E) and management (Culture M), I return to chronological career order and look at the students who enter an engineering school intending to graduate and become employed as young engineers. Largely with no *a priori* view of engineering as a profession, they go to their first classes with high anticipation, reasonably expecting that they are now on course to become *engineers*. What they usually find on offer, surprisingly, is a markedly different culture, namely the culture of *applied science* (Culture AS).

As described in Chapter 2, although these students have chosen to spend their career in Culture E, they find themselves immersed in Culture S instead, which most students almost certainly (and mistakenly) take to *be* Culture E (because they think of themselves as engineering students and their degrees are advertised as being engineering degrees). My author's burden in Part I is twofold: first, to raise the consciousness of engineering students with respect to the world outside the academy and to help them realize that they are being introduced largely into Culture AS, not Culture E, and, second, to make readers see that one inseparable component of the engineering profession—namely, *management*, a

xiii

third culture—can itself be, for some students, broadened into an exciting and fulfilling career.

Part I is intended for engineering students and should be read as early as possible in engineering school. If they are to navigate successfully from applied science to management—two cultures distant—they had better get as early a start as possible. I am committed to the hope that these student readers will be brought to recognize a particular vista of careers that they will not hear about seriously in most engineering schools. That "vista" is the wide variety of careers in *management* for which a background in applied science and/or engineering prepares them very well.

I happen to love engineering students. Not just because, 48 years ago, I began to be one, but because I have lived my professional life with them ever since, as a teacher, as a research advisor, as an entrepreneur, as an employer, as a mentor, as a Program Director, and as a colleague. This is the primary reason I spent my first year of "retirement" writing this book. Students in most engineering schools are being presented with a list of career opportunities that is somewhat biased and truncated. Their teachers (professors) are rarely engineers by any sensible definition, much less managers. Most engineering schools rarely hire actual engineers to teach or advise. (Reason? They are from a different culture!) Similarly, most professors in engineering schools would not be hired were they to apply for equally senior (i.e., equally salaried) positions as "real" engineers in the "real" world.

Professors in most engineering schools are almost always *applied scientists*—a genetically related but quite distinct species. Having themselves migrated the short distance from graduate student to faculty member, the applied science culture of their own thesis supervisor (and likely of their whole department) has been imprinted upon them and they will now try to pass these cultural genes on to all their students. All teachers strive

Preface

to replicate themselves, and the applied scientists in engineering academia are no exception.

There is nothing wrong *per se* with engineering schools training applied scientists. Certainly society desperately needs these innovative young people, and engineering schools with their current staff complements of applied scientists are in a better position to educate the next generation of applied scientists (whose value to society is enormous) than anyone else. The problem is not that applied science students are being trained by applied science professors. The problem, instead, is this: Who is available to train the 80% of engineering students who will not be pursuing advanced applied science degrees, but who want to graduate as engineers? And who is available to inspire the perhaps 10% of engineering students who wish eventually to become senior managers?

Chapter 3 argues that it is the duty of an engineering school to acquaint all of its students not just with careers in civil, chemical and electrical engineering, etc., but about careers in engineering management as well—and to devote an appropriate fraction of its financial and human resources to discharge this duty. All but the smallest engineering schools should develop an "engineering management" program (if, as is likely, they do not already have one) as a high-priority innovation, staffed by teachers who have experience as engineers or managers.

A typical engineering (i.e., applied science) program typically spends an enormous fraction of its time on topics upstream of engineering (such as mathematics and science), and then a few courses on actual engineering, with anything that can be mathematically modeled shown great favoritism. Chapter 4 shows, in abridged form, the entire journey from the most abstract of mathematics to the realities of commerce. Also featured in Part I of this book are two subjects (discussed in Chapters 5 and 6) that are crucial for a future in management, yet are rarely considered in a typical undergraduate applied science education: marketing and office politics.

Engineers Becoming Managers: From the Classroom to the Boardroom

Part II Of This Book

Graduates of engineering schools who have successfully ensconced themselves as functioning engineers in various nonacademic organizations (largely in the private sector) will perceive, at least subconsciously, two tectonic shifts with respect to the culture in which they were engrossed in engineering school. First, although mathematics and science are still recognized as essential underpinnings of their craft, the daily emphasis on these disciplines will be noticeably attenuated. Many other skills, very intrinsic to the practice of engineering but less remarked upon in engineering school, will now be regarded as high priorities. At this point—meaning from many months to a few years after graduation—these young engineers will begin to suspect that the culture of a modern engineering school does not, in some important respects, prepare them for the realities of engineering.

There are many examples of such realities, but the one central to this book is the practice of management. Since it is impossible to avoid some degree of management as one's engineering career develops, one's success will depend on how one responds to that fact. Initially, the word "management" tends to be avoided; words from "management light" will be employed instead, like "supervise juniors," "modify procedures," "participate in planning," "coordinate activities," etc., but eventually this vocabulary will be replaced with words like "makes independent decisions," "long-range planning," "supervise large groups," "determines basic operating policies," "budget responsibility," etc. These words clearly describe *management* activities, and at a reasonably senior level. The next steps—should one have both the desire and the talent to take them—would be in corporate leadership on a senior management team, or in the executive suite, or on the board of directors. Part II of this book is intended for young practicing engineers and should be read as early as possible after graduation.

None of the details of this transition from engineering (Culture E) to management (Culture M) are as important as a frank recognition of the issue itself. It deserves the most careful thought. The problem-solving ability for which engineers are justly famous must be brought to bear upon it. One must decide what the future options and opportunities are, what one's strengths and weaknesses are, and what one most enjoys doing—not just over the next year or two, but over the remainder of one's career.

The solution to this puzzle will obviously be individual-dependent, company-dependent, and even economic-sector-dependent. Part II of this book treats four relevant areas, chosen based on two criteria: first, without understanding these areas it is unlikely that the next stage of career transition—from engineer to engineering manager—will be successful; and, second, they are areas in which engineers, in particular, should be able to excel.

The first of these (Chapter 7) is *risk management*, for whose appreciation the training in probability and statistics that most engineering students get should lighten the load enormously. No business can be successful without planning, and planning requires making assumptions about the future. To achieve the desired (well-considered, well-calculated) rewards requires a commitment to the associated (well-considered, well-calculated) risks.

The second area examined (Chapter 8) is *accountancy*. Anyone who does not understand the relation between his activities and the financial needs of the business (or considers this relationship someone else's problem) is in a self-limiting career. Accounting is not about simple arithmetic. It is about taking proper measurements for internal decisions and about the rules for external reporting.

The third area (Chapter 9) should be a source of excitement for engineers. Their backgrounds and aptitudes prepare them especially well for *innovation* (although innovation is not limited to the technological genres). The relationship of R&D to innovation and the roles of incubators, technology clusters and university laboratories are also discussed.

Finally, in Chapter 10, we examine the important concept of *intellectual capital*, an idea that has been growing in importance ever since someone in antiquity tied their primitive plough to an ox. *Knowledge-based* companies—the ones that do not rely just on finding and exporting resources, or selling business-as-usual products—are heavily dependent on what their employees know, how these employees share this knowledge with other employees in the company, and how all this knowledge gets used wisely and effectively to pursue company goals. Once again, the backgrounds and inclinations of engineers should give them a head start towards internalizing this critically important conceptualization.

Part III Of This Book
I have written the last three chapters for readers whose original degree was in engineering but who are completely committed to a career of management and corporate leadership at the highest possible level. These individuals already have a rich hardcover literature to draw upon—unlike the target readers for Part II, whose books to read are few and far between, and certainly unlike the intended readership of Part I, whose books on this topic are virtually nil. Three chapters are presented that should be of help for engineers who are well advanced in their careers.

The first topic (Chapter 11), *leadership*, shows that, while engineering-based talents are very helpful, they are far from sufficient in making the transition from "engineering management" to full-blown corporate leadership. As at every previous stage in the career paths sketched in this book, there is nothing wrong with deciding that, for whatever reason, senior management is not for you. Perhaps you don't want to; perhaps you shouldn't; perhaps you can't. Fine. The tragedy would be if you *could have*, and *would have wanted* to, but *didn't*, because of a complete focus on day-to-day activities that precluded a proper strategic evaluation of your career possibilities.

The second topic (Chapter 12), on *entrepreneurship*, has good news (you will automatically be at the top of your fledgling business) but also bad news (you are almost singularly responsible for the nourishment and growth of this promising new corporate child). The chapter gives concrete criteria for whether this step is right for you, and also several typical entrepreneurial weaknesses that you should strive to avoid. One of the central problems of growing startups—namely, raising capital—is given special attention.

Finally, in Chapter 13, there is a discussion concerning what is, in some respects, the highest form of corporate leadership—serving on a board of directors. There are lots of lawyers and accountants on such boards, and for good reason, but there are too few engineers—especially for companies that rely heavily on advanced technology either for their internal operations or for their products and services (or both). Part of the problem is that too few engineers migrate into management (and thus the population from which such engineer-directors could be chosen is too thin). It is my hope that this book will make a contribution towards resolving this shortage, by promoting a *management career option*, from as early as the classroom, to as far as the boardroom.

Personal Acknowledgments
Many people have contributed to this book, either to the ideas expressed or to expressing them well. In the former category are Stephen Sorocky (CEO, Dynacon Inc.), from whose many conversations the author was able to examine and sharpen practical management concepts, and Susan Ludwig (formerly Manager, Management Experience Year, University of Toronto), whose understanding of the challenges faced by young engineers making the transition to a management career was very helpful. Neeraj Ghai, who wrote drafts of parts of Chapters 6 and 11 (and who is introduced at the beginning of Chapter 6) also read the entire manuscript and provided useful suggestions.

On the editing side, sister-in-law Helen Curlook (late of the English Department at Laurentian University) contributed her uncanny eye for finding endless typos, while daughter Darlene

Hughes (M.A. in English from Dalhousie University, and Founder and President of IBIS Consulting Group, Inc.) was the book's editor. Lee Födi (Manager of Multimedia Development & Design at IBIS) redrew the figures and prepared the layout and book design.

The author is grateful to the Xlibris Corporation, particularly to Joe Tomines, for printing and marketing this book.

Last, and most important, my sincere thanks to my wife of 48 years, Joanne, whose personal support and interest (and willingness to make sacrifices) sustained me while I wrote this book.

Peter C. Hughes
Aurora, Ontario, 2006

CHAPTER 1

This Book's Motivations and Goals

We Engage the Subject

Engineers, as they proceed through their working life, invariably acquire distinct attitudes toward the subject of "management." Some engineers welcome this intellectual encounter; others are troubled by it; but none can avoid it. The bedrock purpose of this book is to support all these individuals as they develop their personal perspective on management in general, and more particularly on whether and to what extent they wish to embrace management as a career goal. This book seeks to assist engineers, technologists and applied scientists with their career planning, opportunities and decisions, by providing the kind of information that the author believes important and relevant specifically to some form of management as a professional option.

1.1 ONE SIZE DOES NOT FIT ALL

It should be clearly understood that the intent herein is emphatically not to chart some magic career path that is optimal for all engineers. It would be foolish to pretend that there is a yellow brick road for all to follow. In fact it is precisely the opposite viewpoint that will energize our discussion.

To pretend that all engineers are, with at most minor variations, traveling in a one-dimensional career tube, from the time of entrance into engineering school to the point of eventual retirement, is exactly the mindset that this book is intended to expose and dislodge. One's best career plan depends on one's personal attributes and on myriad external circumstances. And, although the subject of this book is how engineers, at various stages of their careers, might view management functions as a career-development alternative, no claim will be made that all engineers should become managers; indeed, no flat recommendation should be made that all engineers should engage in even the most modest administrative activities.

Fortunately, engineering is an outstanding platform from which to launch a wide variety of diverse careers.[1] Thus, even for engineers who venture into non-engineering areas, management is far from the only alternative. Virtually every other profession, trade or specialty is also a theoretical possibility. Some engineers become lawyers; some renovate themselves as physicians; and some undoubtedly have become cellists. This book, however, is confined to engineers interested in management.

1.2 THE ENGINEERING-MANAGEMENT CONNECTION

There is a strong relationship between engineering and management—one that is not tenuous, but robust. Indeed, if one peruses the definitions of *engineering* and *management*, one cannot escape the conclusion that engineering *inherently includes* a significant management component.

> **Fundamental Viewpoint No. 1**
> *Management* functions are essential elements of *engineering*.

---

[1] §2.3 will discuss this good fortune in more detail.

Applied science activities that are sufficiently large, complex, or important to be called "engineering" must, to be successful, be suffused with elements of management.

We can use any of several investigative principles to prove our contention that engineering and management share an important intersection of activities—a major intersection that should banish all doubts. One such path is to examine in some detail the sorts of activities that are carried out under common rubrics like engineering practice, engineering projects, engineering firms, etc. This will uncover much evidence.

However, even this track is too limiting. It assumes that any person whose first degree is in engineering must forever act largely as a professional engineer; else this person is off-track and pursuing some sort of anomalous professional growth pattern. Like monasticism, perhaps, engineering is a calling? Our discussion in this book categorically rejects this view.

> **Fundamental Viewpoint No. 2**
> The study or practice of engineering may lead to many subsequent career paths. One of these subsequent paths is management. Not just *engineering* management as might be inferred from Fundamental Viewpoint No. 1, but *management per se*. This means that management—complex, challenging, responsible leadership roles in many types of organization, and at the highest levels—is a realistic career goal for some engineers.

Engineers Whose Careers Are Dynamic

Lest the reader conjecture that these fundamental viewpoints are merely the peculiar opinions of the author, let us look at what engineers actually do with respect to their careers. For example, Fig 1.1 shows the entrance class to the MBA Program at the business school with which the author is most familiar.[2]

---

[2] The Rotman School of Management at the University of Toronto.

Figure 1.1: Sources of MBA Students (by first degree).

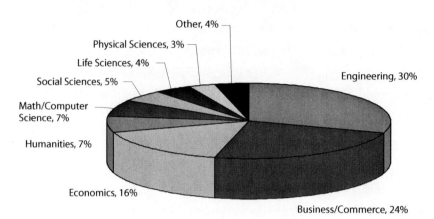

The precise mix of entering students will obviously vary from one business school to another, but it is reasonable to take the data in Fig 1.1 as representative. The key finding is this: A student with a background in engineering is more likely to pursue an MBA than one from any other undergraduate field of study—including business/commerce! Given the steep price tag on a modern MBA from a strong school, these engineers must feel that formal management training is an essential step in their career progression.

What Engineers Actually Do

If one examines what it is that engineers actually spend their time doing, it is not surprising that so many wish to upgrade their management skills. A national survey (1997) of the Canadian engineering profession asked, among other things, that each participant review a list of functions and check off those items on which they spent at least 25% of their time.

The top ten functions from this exercise are shown down the vertical axis in Fig 1.2, with the abscissa indicating the percentage of respondents who had checked off this function. It is clear that many of the ~100,000 respondents listed functions—project management, management/administration, project planning, operations and production, and marketing and sales—that are arguably engineering management functions, while the remaining functions could be called engineering in the narrower sense.

Figure 1.2: The Top Ten Job Functions Reported by Professional Engineers.

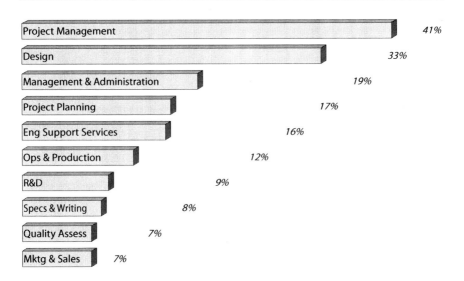

When Fig 1.2 is collapsed into these two more all-encompassing categories, one arrives at Fig 1.3.

Figure 1.3: Percentage of Professional Engineers Who Spend at Least 25% of Their Time on "Engineering Management."

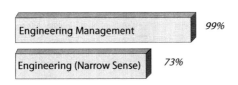

It is starkly evident that virtually all engineers spend a significant fraction (≥25%) of their time doing something they call "management." Moreover, this is a higher fraction than those who carry out strictly "engineering" functions in the narrow sense.

Management Tends To Come With Age

The trend is that engineers, as they grow older, often evolve into managers. This age dependence is apparent (Fig 1.4) from the data collected in the 2002 National (Canadian) Survey of Professional Engineers.

Figure 1.4: Percentage of Professional Engineers Who Have Had, as they Grow Older, Some Managerial Experience.

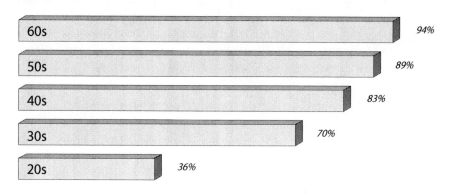

There are likely several reasons for this trend, foremost among which are those shown in the box below. With respect to Point #3 (on compensation), nothing here should be taken to suggest that it would be a good idea were all engineers to become managers just because, on average, managers are compensated[3] at a higher level than are engineers. Engineers who become outstanding designers should be compensated as outstanding designers. Engineers who develop into outstanding applied scientists or engineering researchers should also be paid accordingly.

---

[3]The term *compensation* is preferred here to the more pedestrian *salary* because compensation includes, in addition to **(a)** salary, also **(b)** benefits; **(c)** short-term incentives (goal-determined bonuses); **(d)** long-term incentives (for example, stock options); and **(e)** many other conditions of employment, both contractual and cultural. These matters will be revisited in Chapter 10.

Chapter 1: This Book's Motivation and Goals

> **Reasons Why Engineers
> Tend To Become Managers As They Grow Older**
>
> 1. *Engineering students are often given no clue about how intrinsic management is to engineering, nor about how exciting management can be, and thus come sometimes too late to this realization.* Part I of this book is dedicated to amending these deficiencies for its engineering student readers.
>
> 2. *Management is one of the most difficult functions to do well, and so takes longer to learn.*
>
> 3. *Managers tend to be compensated better than engineers; thus management seems, to many engineers, like the best career path.* This is not a conspiracy, but simply a consequence of Point #2. See also, however, the further comments on compensation just below.
>
> 4. *Many engineers, as the years go by, long for new challenges and broader perspectives.* Management provides these.

This is a multidimensional (or *matrix*) issue. There should always be two overriding principles at work: first, professional engineers should be compensated in proportion to their contribution (of whatever kind) to the creation of value in their organization; and, second, professional engineers should always strive to improve themselves so as to create higher value for the organization that employs them—or for the organization with which they seek to find employment in the future (including, possibly, their own startup organization).

Skills Engineers Believe They Need

Yet another path to demonstrating the robust inextricability of management with respect to engineering would be to ask practicing engineers which talents they feel are most lacking in their armamentarium of professional skills with a view to continued career development. Fortunately, these data are also available[4] from the 2002 National (Canadian) Survey of Professional Engineers. In answer to the question, "Are there any skills that you do not currently have, but that you believe you should acquire to maximize your success in your occupation," respondents replied as shown in Fig 1.5.[5]

Figure 1.5: Top Ten Skills Identified as "Needed" by at Least 10% of Engineers.

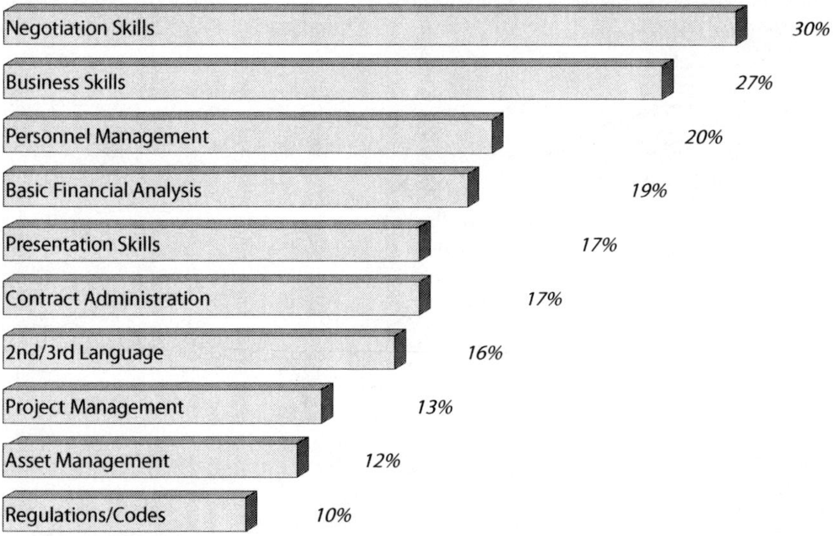

Skill	Percentage
Negotiation Skills	30%
Business Skills	27%
Personnel Management	20%
Basic Financial Analysis	19%
Presentation Skills	17%
Contract Administration	17%
2nd/3rd Language	16%
Project Management	13%
Asset Management	12%
Regulations/Codes	10%

Once again, the incidence of skills that are actually management skills are conspicuous in both number and frequency of identification. These include: negotiation skills, business skills, personnel management, basic financial analysis, contract administration, project management, and asset management. Only "regulations and codes" is more an engineering skill (although one

---

[4] Once again, Canadian data will be taken as representative of similarly developed jurisdictions. The argument here does not rest on fine distinctions, but on the ubiquitous evidence of an intense connection.
[5] Only skills identified by at least 10% of respondents are included in this chart.

that would much benefit from a legal context, and with which, like so many other skills and activities, management should also be familiar); the "second language" skill is neither intrinsically engineering nor management and is probably geographically sensitive. "Presentation skills" is an essential component of any knowledge-based profession, whether engineering, management, or otherwise.

If one were to project Fig 1.5 down onto the dimensions of engineering management and engineering (narrow sense)—for a similar process, compare Figs. 1.2 and 1.3—leaving "second language" to "other" and realizing that "presentation skills" are needed by engineers, managers, and all others who are expected to play a serious part in their organization, one is again forced to conclude that most professional engineers become aware, as they grow older, of a void in their background and training: They are expected to be managers, at least to some minimal extent, yet they are lacking preparation in precisely this professional area.

And thus we have come full circle. This section (§1.2) began by noting the large number of engineers who feel the need for further formal training in management, as evidenced by their widespread enrollment in MBA programs. Now we see why. Most engineers spend a large portion of their time "doing management," a situation that only increases with age, and yet this area has been largely neglected in their engineering education.

1.3 THE THREE PILLARS OF THE ENGINEERING PROFESSION

At the risk of oversimplification, it might be said that, based on the foregoing discussion, the three fundamental skill sets needed to accomplish an engineering task are **(a)** applied science (including applied mathematics), **(b)** engineering design, and **(c)** management, as shown in Fig 1.6. For the most part, engineering schools do a fine job of **(a)**, a fair job of **(b)**, but are quite weak on **(c)**. Other skills, such as effective oral and written communication, or professional decorum and attitude, are important and should be taught; but these are hardly unique to engineering.

Figure 1.6: The Three Pillars of Professional Engineering.

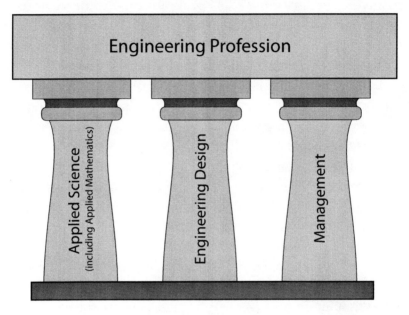

Because engineering is a profession whose creations can be dangerous if not well done, governments and licensing bodies require engineers to "safeguard life, health, property and the public welfare." This implies important *legal* aspects of professional engineering. However, since these are not likely to be career-deciders, they are not shown in Fig 1.6 and we shall not discuss them further here.

To recapitulate: Fig 1.6. is an idealization and does not capture all the necessary elements of engineering. Certainly any specialist will quickly spot missing parts that are important. Nevertheless, all three of the pillars in Fig 1.6 are major and essential, and any attempt to remove one of them leaves the profession quite insupportable. For the purposes of this book, these are the three pillars that deserve clear recognition.

Chapter 1: This Book's Motivation and Goals

1.4 THIS BOOK WILL HELP THREE GROUPS OF READERS

This book has been written for engineers and engineers-in-training. It is aimed at the whole career path, from entry into engineering school to retirement; however, for simplicity, there will be three canonical scenarios (each represented by a major Part of the material) as shown in the box below.

**The Three Target Career Scenarios
(Corresponding to Parts, I, II and III of This Book)**

Scenario 1:
The reader has just entered engineering school and wishes to make wise course choices based on informed career alternatives.

Scenario 2:
The reader is in the early days of his or her career as a professional engineer and wishes to sort out the missing skills needed for, and his personal reaction to, the evolution toward engineering management.

Scenario 3:
The reader is an experienced engineering manager and wishes to consider a career in senior management.

Although these scenarios provide a rational organizing principle for this book, there are two caveats that must immediately be made. First, the time subdivisions are obviously somewhat arbitrary. The epoch of graduation from an engineering school surely creates a natural milestone—although, even here, the possibilities of graduate degrees and co-op programs cloud the timing. As to the exact point in time at which one should stop being concerned with just engineering (in the narrow sense) and begin to reflect on increasing one's career emphasis on engineering management—and similarly as to the time at which one should not think any longer of themselves as an engineer, but as a senior manager[6] or organizational leader—there are no such precise instants.

Which brings us to the second caveat: The career time frame given for each type of reader is the *latest* possible, and certainly never the earliest possible. Thus, one cannot choose university courses and program options later than graduation! Similarly, one cannot expect to have a successful career in engineering management if one spends most of one's early career ignoring, or fighting, that very possibility. Finally, it is extremely unlikely that a reader will ever be entrusted with the controls of an organization's highest office if that reader has not considered, embraced, and planned for this possibility for many years.

To repeat, it is never too early to think about any of the above three career scenarios. However, it can often be too late.

Scenario 1: The Reader Is In Engineering School
No one is claiming that all engineering students should become managers. (Those who are confused about this should read §1.1 again.) What is claimed here is that the "management" subset of career choices should be thoroughly familiar to engineering students and that they should strive to make wise choices from the springboard of engineering school just as they chose engineering while in secondary school.

Unfortunately, engineering students have to swim upstream to accomplish this objective. For one thing, the career assistance available at most engineering schools is concerned primarily with the short-term task of matching the student up with a willing engineering employer—admittedly an urgent and highly useful service.

The greatest obstacle to serious consideration of management as a career flavor is the prejudice of most engineering schools against many actual engineering activities—including management—while stressing applied science and design.

---

[6]The main significance of the distinction between Scenarios 2 and 3 is that one drops "engineering" as a necessary modifier of "manager."

Figure 1.7: Distortion in Much Current Engineering Education.

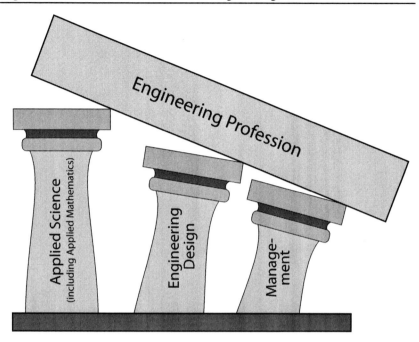

The curriculum and culture of most modern engineering schools looks more like Fig 1.7 than Fig 1.6. This unfortunate bias and the reasons for it are discussed more fully in Chapter 2. In the meantime, consider this situation:

> **Thought Provoker:** Many engineering students are given three full courses in calculus during their undergraduate education—a subject that four out of five graduates *will never use once in their entire careers*—yet they are not taught other skills that they *will use every day in their careers*. What is the explanation for this extraordinary state of affairs?

Most engineering professors—some of whom are not actually designated as professional engineers by the cognizant licensing bodies—are primarily career applied scientists. They are exemplary in the first pillar shown in Fig 1.7, but are not otherwise experienced[7] as engineers.

Scenario 2: The Reader Has Just Started An Engineering Career
Just as the newly enrolled engineering student in Scenario 1 is overwhelmed by the short-term performance goal of doing well in an environment that is much more demanding than secondary school—so as to maintain scholarships and bursaries (or even just pass)—the newly hired engineering graduate in Scenario 2 is overwhelmed by the short-term performance goal of doing well in an environment that is much more demanding than engineering school—so as to maintain career progression (and not be sidelined or dismissed). It is difficult for one to find time and energy for careful consideration of one's longer-term career.

Still, such time must be found, and Part II of this book is intended to assist in this regard. The most important career principle to apprehend is this: When it comes to detailed professional activities (the micro-view) and whole careers (the macro-view), what many assume to be a single path on the pavement is actually myriad trails in the woods. There are many variations, and where one "fits" is continually subject to change. Moreover, the best news is this: You are not merely the target of blind luck, nor are you the pawn in a game you can't either understand or control. Although there will always be a seemingly random component (just as there is with many processes in applied science), it is your responsibility to yourself and to those you care about to spend at least a small percentage of your time devoted to your career research, thinking and planning.

Any employer worth working for is continually monitoring the progress of its employees, seeking strengths and weaknesses, look-

---
[7]The claim is sometimes made that engineering professors who have government or private sector consulting arrangements are thereby experienced in engineering. While these arrangements are to be applauded (see the discussion in §2.4), they are no substitute for full-scale engineering experience.

ing for evidence that, among the hundreds of identifiable skills, the ones needed by each employee to thrive in his or her position are either already present or being developed. More exciting is this realization: If you show impressive skills in areas where help is needed in your organization, you will likely be invited to take on those challenges. In other words, you will not only be good at what you do; you will be becoming good at what you are going to be doing. Your career will be evolving.

The truth of the above three paragraphs is not exclusive to engineering, but can be said of any new position. For professional engineers in particular, however, some of the thought processes that should be helpful are contemplated in Part II of this book.

Scenario 3: The Reader Is An Experienced Engineering Manager
Finally, we treat in Part III the case of an individual who has eyes set on more general or advanced management. This person typically has been performing management or administrative functions for several years within an engineering setting. Many such persons are either doing, or are assisting with, engineering project management. Indeed, project management has many of the essential ingredients present in senior management and therefore provides an important stepping stone for engineers who aspire to more senior positions.

Project management is not the only path to senior management, however. Another path, for example, is to become an entrepreneur and create a growing startup. (This path sounds easy, but don't be deceived.) These and similar issues will be dealt with in Part III.

---

[8]The word "administrative" is often used to mean routine or not-quite-real management. The word "manager" implies at least some leadership, whereas "administrative" implies that most activities are performed within the confines of detailed organizational policies. Many administrative positions have been eliminated in recent times because computers can do many of them more efficiently. The primary interest in this book is in management functions—whatever they may be called—as performed, and performed well, by engineers.

1.5 SUMMARY OF BOOK STRUCTURE

Figure 1.8: Arrangement of Material in the Remainder of This Book.

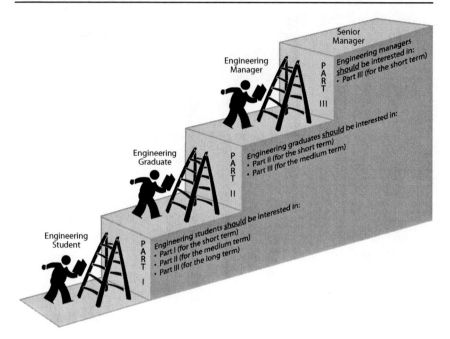

The remainder of this book will be arranged in terms of the three scenarios just described. This is depicted in Fig 1.8.

Part I is primarily for engineering students—although, since everyone in Parts II and III was once an engineering student, they may also be interested in scanning the material in Part I.

Part II is primarily for young engineers who have at least a few years of experience on the job, and who realize that they must now, if they haven't already, address the "management imperative" in their career.

Finally, Part III is primarily for engineers who have already climbed a number of rungs on the management ladder and are interested in broadening their careers to more senior portfolios that are no longer about engineering, although engineering training and experience will always be beneficial.

PART I

For Engineering Students

Should Management be Part of Your Career Path?

If you are an engineering student it is important to be aware that, as you begin to penetrate the engineering world, there are both advantages and disadvantages to formal study in a university or college setting. The curriculum will be well-organized; there will be a variety of engineering specialties available; and the evaluation of progress will be professional and realistic. These are clearly beneficial attributes.

Among the disadvantages will be that certain approaches to an engineering career will be stressed more than others. Greatest stress, from entrance to graduation, will likely be placed on the possibility of graduate research, leading to a career in applied science research—which will be called engineering research. Design, and many other skills used constantly by engineers, will be given shorter shrift. One particular family of careers—the ones highlighted in this book—are rarely mentioned: management careers (Chapter 2).

The thrust of Chapter 3 is to make engineering students aware of these highly attractive career alternatives and to guide them on how they might arrange their course choices and career investigations accordingly.

The focus in Chapter 4 is to show that engineering is one stage in a very long process (the Long Path), from fundamental mathematics, through the recent development of the scientific method, and onward to many other professional specialties that lead to exciting outcomes for society: jobs, productivity gains, and wealth creation. The latter require the energetic activity of professional managers, and engineers are well-positioned to guide this process.

Although mathematics and applied science are largely introspective disciplines, the final goals of the Long Path demand an outward-looking frame of mind, where the importance of the final customer is stressed (Chapter 5). This first Part of the book is concluded by opening a Pandora's Box of potential problems far removed from the calculus course: Chapter 6 explores office politics. Although this is not customarily thought an appropriate subject for discussion in engineering lectures, one's response to these realities may well determine one's gut reaction to management career possibilities.

CHAPTER 2

Applied-Science Bias in Engineering School

Engineering is More than Bessel Functions

As engineering students proceed through their undergraduate programs, typically four years long, they begin to contemplate their leave-taking on graduation, and to envision taking their proud place in the working world as professional engineers. In their senior years, however, they slowly become aware—if they were earlier unaware—of another possibility: they could stay in school longer and study engineering further. Through this effort they could then earn what they believe is an additional engineering degree. Many decide that, if one engineering degree is good, two would be better.

All is not quite as it seems. For several reasons, there is a bias within engineering schools in favor of a career in research—specifically, in what is usually called "engineering research," but what is, in most cases, "applied science (including applied mathematics) research." While this career path of additional study is indeed a noble one, and not to be denigrated by those for whom it is suitable, it is equally necessary to observe that it is not a suitable path for all—or even most—young engineering students. This basic observation is difficult to square with the curricula usually offered by, the advice usually given within, and indeed the entire cultural environment of, a typical modern engineering school—all of which have a heavy emphasis on engineering research.

The motivation for this extensive discussion here is to make young undergraduates aware of this bias, and so to assist them with more informed career decisions. Virtually nothing said in the rest of this book (and especially in this Part I) will make much sense to a student unless he or she is aware of this strong bias.

2.1 FURTHER DESCRIPTION OF THE BIAS

Most engineering professors and teachers do not see the partialities just mentioned above as being biases and indeed most are not even aware of them.[1] They simply have internalized an ideology and have devoted their professional lives to that ideology. The latent prejudice is that the highest calling of an engineer is to operate as a researcher, more specifically, to conduct research into applied science (while using applied mathematics as much as possible). For strategic reasons, the latter enterprise is called "engineering research" within an engineering school. Indeed, a nontrivial fraction of the research conducted in engineering schools may arguably *be* engineering research, there being no sharp boundary between applied science and engineering.

Status Among Possible Careers

Many professors (including engineering professors) regard their profession as a secular "calling," literally as a "vocation" in the original meaning of that word—based on the same word (from the Latin) used by monks and prelates. It follows that the job is, as they

---

[1] This may be partly due to the fact that (in the words of Noreen Calderbank, manager of licensing for Ontario's professional engineering association, as quoted in *Engineering Dimensions*, May/June 2005), "We discovered that a lot of engineering professors aren't licensed...," and further that "...some unlicensed engineering professors, many of whom have been teaching for many years, may take exception to having to take additional courses or write examinations to qualify for a P.Eng." One wonders whether medical doctors are being graduated after instruction by professors "a lot" of whom aren't licensed physicians. The author confesses to not knowing whether this is some sort of local phenomenon, but suspects it is not an isolated one.

Chapter 2: Applied-Science Bias in Engineering School

say, 24-7-365. Granted, all professionals, by the nature of their roles, do not work by the clock as unionized workers do. They do not "punch in" and "punch out," nor do they normally get any direct compensation for overtime worked.

Nevertheless, most in the professoriate take matters much further, believing that they are professors even when asleep or at the cottage. Many feel most comfortable, even long after retirement, continuing to work at the office as researchers in their field, though the paychecks have long since dried up.

Figure 2.1: The Career Status Hierarchy, as Assumed in Most Engineering Schools.

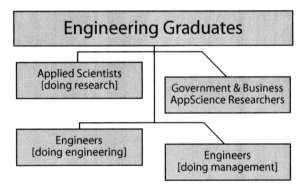

This world view is illustrated in Fig 2.1, which places these professors, who are primarily applied scientists (and who are, in engineering schools, simply called "engineering researchers" or "engineering professors"), at the top of the pecking order. It is also recognized that applied science research is conducted at institutions other than academic ones—principally in government laboratories and high-tech businesses.

Students who are not permitted (or who choose not) to study further towards advanced engineering degrees should realize that the world of engineering research is largely inaccessible to them as a future career. While these students are certainly not referred to as "dropouts," and are not considered "failures," their consignment to becoming (merely?) engineers tends to be tinged with some sadness, and their career paths are seen as somewhat sub-optimal, especially if they have demonstrated strongly the skills needed for applied science research.

As depicted in Fig 2.1, students leaving school after a BASc (or equivalent) almost invariably enter the profession of engineering. At this point, the second phase of the bias trickles in. If they can't be applied scientists, these graduating engineers should at least strive to become better engineers, as part of life-long learning. Improving oneself is clearly desirable; however, the bias becomes apparent in that, if one subsequently undertakes professional development aimed not just at improving engineering skills, but management skills as well, this choice tends to be viewed as leaving the fold, perhaps even joining the dark side.[2] The stance taken in this book is exactly the opposite. As shown in Chapter 1, management is incontestably part, and a critically important part, of the engineering profession. It is the third pillar (Fig 1.6). Moreover, the career path for young engineers should not be limited to engineering management (an admittedly vague term); indeed, it should not be limited, period.

One Pillar Is Weak And Neglected

Thus, the pillars of engineering depicted in Fig 1.6 become, in the ideological environment of the typical modern engineering school, and as presented to young engineering students, distorted in the fashion shown in Fig 1.7. The emphasis of core curriculum is quite discernibly

---

[2]The author's first (but not last) experience with this phenomenon was 35 years ago. Asked temporarily to fill a summer position in middle engineering management at a high-tech company, he was disturbed by the attitude of the engineering staff toward their managers. Regarding managers who had engineering degrees, the question was, "Why are they spending time on unimportant matters (like, presumably, budget, schedule, and preparing for critical design reviews with the customer) when they could instead be helping out with these difficult engineering problems they have assigned us?" Regarding managers who did *not* hold degrees in engineering, the question was, "How can these non-engineers pretend to be competent at managing engineers?" When the above two questions are taken in tandem, they imply that the set of paths to engineering management is the null set! Clearly, to these folks, engineers should not become managers—not even *engineering* managers.

on applied science, with some opportunity to develop design skills. However, when it comes to the engineering-management nexus, there is at best a neglect, or at worst an almost palpable dislike, in evidence.

Many BASc courses that initially appear to have at least some management content, with labels such as Engineering Economics, turn out to be much more about engineering than about economics. Examination of the course content swiftly reveals the underlying assumption that the course aim is to make the engineer a better engineer—a worthy goal, of course. Still, it is rare that such courses aim better to prepare the student to ascend the ladder of leadership within their future organizations.

2.2 REASONS FOR THE BIAS

The bias under present discussion is not a malevolent one. The reasons for its existence are simply the characteristics of any group of human beings who operate at a high performance level within a relatively closed environment.

Nevertheless, this bias can have unseemly consequences for the broad curriculum of an engineering school, and—more to the point for the aims of this book—can impart to young students a distorted set of professional values and an assessment of career opportunities that is sadly truncated. There are several related reasons for the bias.

Reason 1: The Urge To Clone Oneself

Experience in human affairs indicates that, unless there are countervailing forces, teachers in any field will tend to arrange matters—including their conversations with students, their course material, their exam questions, and their subtle or not-so-subtle bits of advice—to produce graduates that are clones of themselves. The ideal undergraduate student, from the professor's standpoint, is one who is very talented in mathematics and applied science and who would make a fine addition to his or her research group on completion of the BASc.

During a further period of "graduate research" (to use North American terminology, or "*postgraduate* research," to use British terminology), the student is further trained as an applied scientist and researcher. This process is advertised to take only a "few years," but in practice often takes 6–10 years. It is an enormous investment of the student's time (up to 25% of his working life) and of his money (salaries foregone can easily reach a third of a million dollars, not to mention student debt).

If the student is on the best path, these vast investments and sacrifices are advisable and unavoidable. However, if this path has been taken simply because another path that was a better fit for this student was not made visible, or even actively concealed, this state of affairs is obviously regrettable.

Reason 2: It's Hard To Lose Your Best

No organization enjoys losing its star performers, and although educational institutions by their very nature have a complete graduating class of departees every year, ways can be found to convince the best and brightest (as defined by the prevailing culture shown in Fig 2.1) that they should stay. Many professors, showing marketing skills never learned in any classroom, find occasions to explain to their undergraduate engineering students how such an "advanced degree" would add further luster to their résumés. Wouldn't a second (or even a third) degree give them an edge in the competition for highly respected positions after graduation? Wouldn't further study launch them into just the sort of career they had entered engineering school to find in the first place?

Significantly, the only students who will actually be accepted into these advanced programs of graduate study are precisely those who have out-performed their classmates on a curriculum focusing almost exclusively on mathematics and science. Proficiency in mathematics and science (applied or otherwise) is essentially the primary skill set that is highly valued in the modern academic-engineering culture and is therefore regarded as the most relevant barometer of useful ability. A doctoral thesis that merely designed a

Chapter 2: Applied-Science Bias in Engineering School

new product with a likely new $100M market—but that was not awash with mathematical analyses—would probably encounter tough sledding at the final oral defense. Other students, less blessed with these precious skills, have to settle for . . . an engineering job.

Much of an engineering professor's reputation centers on coaching bright young graduate students into producing publishable research. The reader will not be surprised to learn that the engineering professoriate is exceedingly loath to lose these young potential professors to (what many of them regard as) more pedestrian engineering positions elsewhere (see again Fig 2.1).

Reason 3: Extrinsic Influences

Even if a single leading engineering school, in a fit of clearheadedness, decided to opt out of this overemphasis on engineering research, and to apply an appropriate emphasis on mainline engineering—and even to develop an apt emphasis on engineering management as well—this would be very difficult to achieve because it would exact a considerable price from the engineering school community. Schools are ranked nationally and internationally, and these rankings generally highlight *research* quality and quantity, much more than *teaching* quality and quantity, to arrive at metrics that are recognized and accepted as arguments for the high stature of the school. Good teaching in engineering management would be counter-productive.

Reason 4: What Gets Rewarded Gets Done

Contrary to the humorous anecdotes treasured by students—and told with gusto at graduating parties—bad teaching is no longer tolerated in any good engineering school (if it ever was). Still, teaching is viewed by many professors more as a necessary job requirement than as a passion. The tone of the discussion at promotion and tenure meetings quickly reveals why this is so: teaching performance must be *competent*, but research performance must be *outstanding*.

Virtually all modern engineering professors do make serious attempts at effective teaching, and many are dedicated enough or talented enough to be cheered on by their students. Still, from the professor's standpoint, the research-centered career reward system mandates that any such striving for classroom success be based as much on empathy as on career incentives.

Reason 5: Papers, Sabbaticals, Etc.
As a further example of research bias, when engineering professors make discretionary judgments as to how they will spend their professional development time, they choose, almost without exception, to go to conferences that feature *research* papers. Incidentally, presentations at such conferences by representatives from companies showing actual engineering achievements are often disparaged by academics as being just marketing, which is undoubtedly meant to imply something slightly unsavory.

Travel expenses to a conference are generally not claimable by an engineering professor unless he or she gives a paper. No paper, no go. This algorithm is especially vexing if the go is to Rome, or Hawaii, or the Costa del Sol. The general idea is reasonable enough: a research conference is a venue in which researchers in the same field come together to exchange views, methods and results. Who could argue with the benefits of such an exercise?

This is a classic example of short-term benefits vs. long-term benefits, and about balancing competing priorities. An international conference in Hawaii, well utilized and with the world's leading researchers assembled, can have a salutary effect on one's research perspective. The problem is that if this is the *only* type of conference that engineering professors ever attend, or ever want to attend, they will have very little that is inspirational or even instructive with respect to their understanding of engineering and to their effectiveness as teachers.

This is especially poignant when it is realized that, although engineering professors (and professors generally) have spent many years honing their research skills, they have little direct professional training in teaching. Most learn on the job, and from the good (and bad) examples of their own earlier professors.

Chapter 2: Applied-Science Bias in Engineering School

Then there are the sabbaticals—employment leaves with a high percentage of regular financial compensation—intended to re-energize, re-inspire, and re-enthuse engineering professors in their difficult task of being always on the leading edge of their field. These are almost invariably spent in a research environment—most often at another university, but sometimes at a government laboratory—and least often in the research environment of a high-tech firm, much less doing any actual engineering. It is difficult to acquire definitive numbers on this subject, but the number of engineering professors who spend their sabbatical leaves either doing what engineers do, or developing material and skills germane to their teaching activities, represents a small minority.

Reason 6: Prior Selection

The entrance requirements for engineering schools strongly emphasize ability in mathematics and science. Certainly this emphasis is reasonable in a narrow context. These gifts should be highly prized; after all, they represent one of the pillars of professional engineering (see Fig 1.6). Moreover, other relevant abilities—one thinks of design and management, the other two pillars in Fig 1.6—have not yet been developed sufficiently by primary and secondary schooling to enable proper measurement.

Nonetheless, the bias at entrance towards mathematics and science ability is not diluted as additional engineering skills are taught and mastered. Except for some instruction in design, these additional skills are taught rarely if at all.[3] Some students are not even aware that they exist.

2.3 GENERAL BENEFITS OF UNDERGRADUATE ENGINEERING

It should be emphasized at this point that the negative tone of the debate thus far relates only to sins of omission, not of commission.

---

[3] The author has had the joy of knowing a few undergraduate engineering students whose organizational and communication abilities were breathtakingly mature. These were usually based on personal aptitude and/or family experience. All left after the BASc.

The curriculum offered by a typical engineering school is usually of high quality and very helpful for a subset of their students. The problem is that these same curricula are bereft of crucial kinds of career-foundational material for the remainder (likely the majority) of their students. Any career advice offered is also generally biased in favor of research (or even professorial) aspirations, which is simply unrealistic for most engineering students. Nevertheless, one must recognize the benefits of an engineering education, even in its current sub-optimal form. Three of these benefits follow.

Benefit 1: New (Post-secondary) Challenges Are Salutary
It is well known that entering *any* undergraduate program in university is a life-changing experience. Adapting to the greater freedom, the higher standards, and the elevated intellectual level of a good university often pose challenges to these young adults. These changes can be navigated to the extent that the student has the mental reserves, the organizational skills, and the social ability and maturity to manage such a development. Only macro-transitions like birth, marriage, parenthood, divorce, dismissal from work, or death, have a comparable impact on one's psyche.

To achieve most of the benefits of the transition to university life, it doesn't really matter precisely what professional or academic area one has chosen to study, so long as the college or university is a *bona fide* institution. Therefore, although engineering studies at a strong university make an excellent arena for general post-secondary mind-training, the most general benefits of engineering are not unique to engineering but are available from any challenging, coherent undergraduate study program.

Benefit 2: Math Classes Offer Special Benefits
Although, as mentioned clearly above,[4] there are serious limitations to the typical academic curriculum for most engineering students,

---
[4] And as will be further discussed below, in §2.4.

there are also some undeniable benefits as well. The first countervailing aspect is that the frankly outsized list of mathematics course obligations faced by the students enrolled in "engineering" do indeed contain fundamental material.[5] The theories, assumptions, and experimental evidence covered in class are unlikely to change soon, perhaps not in the student's lifetime. The education paid for so dearly by these students and their families will at least have a half-life longer than one energized primarily by the latest fad.

Although mathematics courses are, one suspects, mandated by many engineering departments in order to prepare students for their true calling (as applied scientists), the primary benefit for many (likely most) students is that these straight math courses display the most rigorously logical body of material the student is ever likely to encounter. Subjects like advanced algebra and calculus, though mostly irrelevant for those whose career paths are not of the applied science variety, nevertheless train certain qualities of mind in young engineering students. These qualities include: first, a demand for a precise definition of the problem (if possible); second, a call for an identification of the precise data (if available) that would make the problem soluble; and, third, a commitment to the difficult mental work required to compose a careful, elegant solution to the problem.

Benefit 3: Science Classes (Should) Create Healthy Skepticism
The many science courses required of undergraduate engineering students are extraordinarily profitable to their intellectual development. Especially if the *scientific process itself* is adequately stressed (as more fully discussed in Chapter 4), science courses can provide a life-long approach to the endless stream of claims

---

[5]As a young assistant professor, the author undertook a survey of local aerospace companies to ascertain how, from their viewpoint, the aerospace engineering curriculum at his university might be improved. Some replies were real shockers, and contained no notion of the responsibility to impart strategic knowledge. One reply to the survey implored that we should teach how many rivets should be used to repair a damaged aircraft wing. Presumably, this was the issue-of-the-week for that particular respondent!

to which they will be exposed.⁶ Most of this vacuous material and spurious nonsense will be encountered outside the university, but frankly some of it can be found inside as well.

Science is neither a collection of data, nor a recitation of alleged facts. It is a process of extremely hard work through which truth can be approached (if never quite reached). Unfortunately in some respects, it is generally the *results* of the scientific process that are stressed in engineering school, because it is these results that have practical import, and the more intellectually beneficial consideration of the scientific *process* that led to these results is given shorter shrift.

In brief, mathematics and sciences courses, since they are disciplines based on logical processes—whose results have moved mankind further in the last two centuries than the (perhaps) well-intended but ultimately defective thought cultures of all the centuries that preceded the age of science—create *a magnificent discipline of the mind* for bright young students. This benefit is conferred whether their ultimate role in society will be as applied scientists, researchers, technical innovators, designers, engineers, managers, leaders, citizens—or any other intellectual pursuit.

2.4 NEW CURRICULUM OPTIONS SHOULD BE OFFERED

Curricula vary greatly in engineering schools, both from school to school, and even from one era to another in the same school. Criticisms that may be appropriate for one school at one time may be inappropriate for another school—or even the same school at a different time.⁷ The one challenge that is likely to be spot-on for almost any engi-

---

[6]The student should be expected to challenge **all** assumptions, including the ones learned in class. Science is not an ideology; it is a truth-finding process.

[7]If the reader will again forgive the illustration of a personal experience, the author studied aerospace engineering as his undergraduate specialty. He did not see any part of an aircraft drawn on the blackboard until a senior-year course in Aircraft Design (taught by a prestigious engineer in the local aircraft industry, because there was no one on the faculty able to teach it), and he never saw a single spacecraft drawn on the blackboard, at all, as an undergraduate. (Perhaps the fact that Explorer I entered space the same year the author entered engineering school may explain the latter observation.) Happily, aerospace engineering is now a much more mature curriculum at the author's *alma mater*.

Chapter 2: Applied-Science Bias in Engineering School

neering school at almost any time is the bias, at least to some degree, in favor of applied science and against the other pillars of a complete engineering education (Fig 1.7).

Inbreeding Can't Be Good
It is a remarkably rare event when a practicing engineer or anyone with out-of-academy experience is hired to teach and/or conduct research at an engineering school. This is regrettable because the fresh thinking from the actual engineering profession—not the resident applied scientists—goes missing.

The reason given for this hiring policy, which is tantamount to a boycott against experienced engineers teaching in engineering school, is that new PhD's in "engineering" are generally cheaper to hire. They are also generally inexperienced in real engineering, and in fact are inexperienced in everything except a long graduate-school experience in applied science in a very narrow specialty. They have also never been taught anything about teaching—except by examples, good and bad.

As their career develops, these young applied scientists, who have now spent virtually all their life in academe, join the culture (Fig 2.1) that says, first, "The best engineers are PhD applied scientists," and, second, "The best PhD applied scientists are the professors in engineering departments." The last claim is not too far off, but the first betrays a clear bias and a dangerous distortion. Since a new professor can accept a new student who graduates as little as five or six years later, an academic generation turns over every five or six years. With no outside professional-engineer genes to moderate the culture (gene pool), it takes only a few decades for a rock-solid bias to form. This bias then proves refractory to any different suggestions, including presumably the ones made in this book.

Letting The Light Of The Profession Flood The Cloister
There are many ways to strengthen the connection between the meaning of "engineering" as used in teaching academies and its

meaning in the rest of professional society; some are more effective than others. We deal first with some ideas for helping students by rearranging the priorities of their teachers; other suggestions for helping students more directly will be made in §2.5.

Here are some effective suggestions for action:

> **Policies to Bridge the Inside–Outside Culture Gap—The "A" List**
>
> ***A1: Hire some experienced professional engineers as professors.*** For such individuals, a PhD is not a negative, but is not required. Reliance is placed instead on a résumé (which should not be confused with an academic curriculum vitae) stressing engineering situations faced and contributions made. A proper job description and a proper job interview process would be mandatory. These hires will not be second-class citizens in their departments (though, perhaps, they will have a less-than-stellar familiarity with Bessel functions), and will be used as respected resource persons on curriculum development, thesis selection and direction, and as key teachers of the profession. The requirements for "original research contributions" will be de-emphasized or absent altogether, being already well-represented by the career applied scientists in the department.
>
> ***A2: Ensure that all engineering professors spend time in non-academic positions.*** Summer positions in the private sector, or government, or NGOs, or similar, would be strongly encouraged. Two consecutive sabbaticals would not be taken in a purely academic environment. Other similar measures would be encouraged through effective policies.
>
> ***A3: Redesign professorial performance metrics to include recognition for achievements in the profession.*** Use for salary progression, tenure decisions, and promotions. Not as easy as counting papers or contract dollars, and does require judgment. What gets rewarded gets done.

Chapter 2: Applied-Science Bias in Engineering School

The most fundamental of the above three policy areas is A3; without a substantive adjustment to the reward system, the first two policy initiatives (A1 and A2), and others in the same spirit, are doomed to fail.

Attempts to implement these policies—however energetic such attempts might be and however beneficial the policies might be—will inevitably meet with staunch resistance from the current engineering professoriate. This reaction is not difficult to explain. In addition to the natural dislike all humans have for change (with an intensity roughly proportional, *ceteris paribus*, to the length of time already spent in their established system), the existing faculty members will have optimized their career activities and work priorities to the existing performance (and cultural) metrics and will hardly be receptive to new metrics—even though the results of the changes are beneficial to the shared enterprise. This all requires courage.

Some additional policy changes that would further align the typical engineering school with the profession it purports to represent are as follows:

Policies to Bridge the Inside–Outside Culture Gap—The "B" List

B1: Hold seminars that are of interest to (and perhaps even given by) outside engineers. Find a way to encourage students to attend (e.g., make it part of a course).

B2: Consider turning 10% of research space over to high-tech incubators. But watch ethics carefully (e.g., market prices for all university services used after three years, and out of the incubator altogether after five years) especially if academic staff are significant shareholders.

continued >>>

33

> **B3: Include an engineer on Master's and Doctoral dissertation committees where the subject matter makes this desirable.** If the subject matter makes this only rarely desirable, something is wrong. Explanations (and stronger) may well be necessary if the central university administration insists on applying strictly science paradigms to engineering dissertations, or resists the revolutionary idea that engineering doctorates should have some engineers on the judging panel.
>
> **B4: Many other creative bridge mechanisms.** Once staff and students become excited about building bridges between the school and the profession, countless exciting mechanisms will be found to amplify this movement.

These items (the "B" list) should not be expected to be very effective (or even happen) unless policies like those in the "A" list above are brought enthusiastically into force.

Engineering Professors As Engineering Consultants

We close this section with some brief comments on the role engineering professors should play with respect to outside professional activities—referred to here, for short, as "consulting." The summary answer to the question, "Should professors be permitted—even encouraged—to engage in consulting?" is this: Handled properly, consulting is a win-win-win-win. It is **(a)** a win for students, who will get more appropriate course material and thesis topics (and perhaps the opportunity to participate in these consulting relationships); **(b)** a win for professors, who will experience the excitement and professional development associated with working on real-world problems (and who will be grateful for the opportunity to augment their income though consultancy compensation); **(c)** a win for the school, who will have a teaching and research faculty that is better compensated, more professionally sophisticated, and more appreciated by the outside community; and **(d)** a win for the consultants' customers, who will become more supportive of the role of engineering schools and more receptive to fostering university-business interactions of many types (from seminars to major research contracts to hiring students on graduation).

The point to which we always return is that there is a cultural bias within academe against many of the activities that characterize professional engineering, and that this bias should be urged to disappear, especially as it relates to the student learning experience. If this premise is accepted, it would then be fatuous to discourage professional dealings between engineering faculty members and the engineering community at large.

Person-to-person interactions that are continual and ubiquitous are more valuable than all the high profile agreements and policy statements that can be dreamt up by administrators. Still, there are dangers of excess and potential conflicts of interest. The school's leaders, in addition to encouraging consulting, should ensure that policies are in place to assure that these activities are defensible within ethical guidelines.

2.5 NEW PROGRAM OPTIONS SHOULD BE OFFERED

Many engineering schools are quite large—large enough, in fact, to be called "Faculties." They have the heft and breadth to offer multi-stream alternatives to their undergraduates. The most common streaming principle (what a marketer would call segmentation) is by engineering *specialty*, or *type*, or *"discipline."* The student must choose one of {aerospace, mechanical, mining, materials, industrial, civil, computer, chemical, electrical, etc.} engineering. It may well be that some students would prefer a more general approach, in which they are exposed to *all* the engineering disciplines, but the discipline-based departments of the typical engineering school would have no interest in such a common curriculum.

Still other approaches to the preparations of students for their career aspirations should be developed. In some engineering schools, control by the discipline-based departments is so strong and so historically entrenched that no preponderance of rational argument would ever suffice to produce any "non-discipline" basis for curriculum classification. The following remarks are addressed to students who seek a broader assortment of career preparations, and to engineering schools that are open to new curriculum ideas.

The Applied Science Stream
A glance at Fig 1.1 indicates another segmentation principle, namely, the possibility of streaming engineering students into those most interested, respectively, in applied science; in design; and in management. These will each be considered briefly, in turn.

We claim that there is a bias in favor of applied science in most engineering schools, and that most teachers and professors are (no surprise) from the applied science school of thought. It is natural, therefore, for the strongest schools to develop an "applied science" stream, whose entrants are judged primarily on their aptitude for, and performance in, mathematics and science. These applied science programs in engineering school are often called "engineering" science, to satisfy critics who say that these programs do not contain enough "real" engineering instruction.

This process of helping our best young applied scientists to apply science to practical purposes should require no apology, whether there is any really detailed engineering involved or not. Where, otherwise, will such applied scientists be educated? Can we trust the science departments (usually found in arts faculties) to shoulder this important duty? The author would certainly vote against that alternative: whatever the weaknesses or biases of engineering schools may be, they are much more likely to be supportive of concepts like "practical," "application," "innovation," "commercialization," "business," "corporation," "earnings," "profit," and "wealth creation" than an arts department.

Applied science programs in engineering schools usually cultivate and reward the best applied science education that money can

Chapter 2: Applied-Science Bias in Engineering School

buy. Their graduates typically proceed—usually after much further study in graduate school, and often with a PhD in "engineering"—to a career in technology-motivated advanced research that is of the highest quality. Some will take up positions in government science laboratories; some will leave for large corporations with R&D programs; and some will be hired as young professors. These are the alpha-researchers and fundamental innovators in our technology-hungry society. The story of these individuals, both for themselves and for everybody else, is a very happy one.

The real concern here is not that applied science students in engineering (who brilliantly navigate the interdisciplinary gulf between science and its practical applications) get too many applied science courses, or that they get too few "real engineering" courses; it is precisely the obverse. Students who wish to become—or at any rate eventually do become—"real" engineers get too many applied science courses and two few engineering, design and management courses. The 90% of students who will not eventually become applied scientists should not be subject to a curriculum that is based on the needs of the other 10% of their colleagues. They should benefit from a curriculum that is designed for *them*. A significant subset of these students is interested in, and has great talents for, careers that entail both the design and management pillars of the engineering profession.

The Design Stream

To return to Fig 1.6 (the ideal) and to Fig 1.7 (biases included), it is clear that *design*, being a fundamental pillar of the engineering profession, should accordingly be a fundamental component of one's engineering education. Still, of one thing there may be no doubt: design is fundamentally different from applied science, as it is also from management.

The nature of design is that details are crucial. Thus, for example, it would be bizarre to attempt to construct a strong curriculum that emphasized design independently of the "Engineering Discipline" system of student segmentation. Nevertheless, there are certain systematic approaches to design that are followed, plus or minus, by the best companies. This material qualifies not as "theory" but as "best practices." It would be cruel not to acquaint engineering students with some of these design strategies.

Thus, design should not be a separate "stream" at all. Design should enter the curriculum in two ways. First, every engineering graduate worthy of the name should have had at least one course in the best-practices design approaches mentioned in the last paragraph. Second, for students who are not planning on becoming applied scientists or engineering managers, there should be a significant emphasis on design within each segment of the "Engineering Discipline" system, with the level of detail appropriate to each segment.

Finally, a comment on certain historical practices. So-called "design" courses that teach about the various weights of paper, the sundry thicknesses and darknesses of design pencils, or how and when to erase a line that is too thick or too thin, must be forever banished. If they still exist, they are an insult to engineering students for two reasons: **(a)** if the upward career migration and productivity increases needed by a growing economy are to be achieved, this sort of retro educational material must be jettisoned; and **(b)** design nowadays is performed largely with the aid of design software packages, which any attractive employer will already be using.

The Management Stream

As suggested by Fig 1.7, the attention paid in most engineering schools to the third pillar of engineering—management—is small to nil. This should be vexing to a large number of engineering students in view of the statistics cited in Chapter 1 on how many young engineers become engaged, either *ab initio* or eventually, in some form of management, and who *want* to engage in some management activities. It is difficult to conceive of serious engineering activities being pursued in the absence of any organization, any

Chapter 2: Applied-Science Bias in Engineering School

strategy; any potential or actual customers; any financial considerations or limitations; any partners or colleagues; or any realistic vision. In short, without any management.

Yet this is precisely what a management-free engineering education assumes (and produces). Students are trained to be engineers in the narrow, traditional academic sense. On the contrary, engineering schools that want to train their students for real-world positions must address the management component (pillar) of engineering.[8] It is not good enough to say, "The ones who are interested will learn on the job," or, "If they want they can study later at a management school." Granted, they will learn much "on the job." And granted, engineering schools can spend only a limited fraction of curriculum time on management, but that limit is rarely approached.

Another, more ambitious approach is to develop an extended joint program wherein a stream of engineering students is able to take management courses as well as engineering courses, leading to a commerce or management degree as well as the engineering degree. Joint programs of this nature[9] take somewhat longer to complete, but give students two degrees and many more tools with which to pursue their chosen careers.

The MEng Stream

Some engineering schools have developed MEng degrees—master's degrees in engineering—based largely on course work, not on research. These are intended to prepare students for life as professional engineers. They sometimes run into difficulty when professors (having been selected as the fittest for applied science research) can't or don't want to teach non-research engineering. This can be remedied by appropriate reward policies and hiring initiatives.

---

[8] Not to mention the many young engineers, hopefully many, who aspire eventually to rise to the most senior levels of leadership and responsibility in their organizations.

[9] Other joint programs would also be welcome additions to the curriculum. Engineering/law and engineering/medicine come to mind; however, this book is focused on engineering/management.

It is also refreshing to have a putatively engineering degree called an engineering degree, unlike the BASc (which accurately says that it is in applied science) or the truly odd PhD (which is apparently in philosophy, hardly a synonym for engineering).

Time On The Job
Finally, one of the most effective ways for students to bridge the culture gap between their engineering school and the real world of engineering is by spending several terms in an engineering workplace. Many schools have these "cooperative" programs and they are highly beneficial. There are few students whose horizons are not greatly and positively affected by these apprenticeships. They also frequently get a leg up on employment possibilities if their performance is well received.

CHAPTER 3

Important Decisions in Engineering School

Filling in Some Missing Pieces

In primary and secondary schools, students have ample career counseling materials readily available. There are designated teachers who are committed to assisting students in their career choices, either informally or through a formal vocational guidance office. One of the points these counselors drive home unrelentingly is that it is *never too early* to consider career options. In addition to the obvious matter of course and program selection, students also learn that behaviors, attitudes, and levels of commitment, even at a seemingly early stage, have a strong influence on success, and are wise investments of the students' time. Course and program choices can clearly have ramifications for decades. To state the issue even more positively, well-prepared students will benefit from wise choices, not just for many years, but likely throughout their entire careers. Indeed, it is never too early to consider career options, to make choices (if only tentatively), and to infer the best course decisions.

Most engineering undergraduates know all this and could have written the above paragraph. Not by accident did they decide to choose their challenging profession and not by happenstance did they achieve their high performance in exactly the right courses in secondary school to succeed in the highly competitive entrance process of the modern engineering school. They had decided, likely many years earlier, that engineering was right for them as a postsecondary learning process, and now they have arrived at a new and higher school, determined to give a good account of themselves and to succeed in this next phase of their career.

Yet, sadly, many of these same students tend at this new stage to ignore the wisdom of continually reflecting on their careers, and on making choices that are wise for them. They act as though everything has been decided; the die has been cast; they have chosen engineering, and likely even the brand of engineering (that is, civil, electrical, etc.), and there is nothing more to be done except to graduate and successfully get an engineering job.

3.1 UNDERGRADUATE CAREER CHOICES

Part I of this book is designed to disabuse undergraduates of any such notion. Many career alternatives are still available and some of the most interesting of these require that informed choices be made right away. From among the almost infinite range of choices, this book focuses on careers leading to a management role of one kind or another.

As Chapter 2 and Fig 1.7 have made clear, students at most engineering schools are in fine shape, as far as curriculum and culture is concerned, if their passion is applied science. They will be well educated and, if they work hard and have the royal jelly for creative engineering research, they will go on to advanced degrees in applied science, to form the bedrock foundation for the technological creativity that propels our civilization forward.

If, instead, some students have the gift[1] for design, things won't go quite as smoothly, but they will still find specific courses

---

[1] It is not being suggested, of course, that all students must be whizzes at one of applied science, design, or management, and no more. Some students are rather sharply focused, while others have abilities that span a wider spectrum. And no load of political correctness should obscure the recognition that some engineering students, sublimely gifted, are able to excel in a wide range of disparate pursuits—all with seeming effortlessness (perhaps the true test of genius).

Chapter 3: Important Decisions in Engineering School

and program nuggets that nurse their creative endowments to a fulfilling future. Although some of these individuals will have to endure a highly intense maths and science curriculum that is not especially tuned to their mindsets, most will persevere eventually to create engineering designs that are useful, novel, esthetically pleasing, and profitable. Modern design software will amplify their creative output and they will be in high demand by growing companies.

If, finally, some students like to put things together rather than take them apart; if they welcome the opportunity to work with others in a team environment where each member is challenged to do their best for the good of the enterprise (rather than nestling away to think their very own technical thoughts); if they want to look at the forest and not just at Leaf #1 on Branch #2 on Tree #3 in Clump #4—in short, if they are interested in the broad picture and playing a leadership role in painting it—these students, who may be management material, are generally at a marked disadvantage in most engineering schools.

Program Segmentation—The Ideal Curriculum Structure
The word "segmentation" represents a key concept in marketing; it will be explained in Chapter 5. Briefly, it is a process wherein one breaks one's customers into identifiable subgroups, based on some discriminating[2] principle, such as age, sex, etc. The chief segmentation principle in engineering schools is based on the engineering type, or discipline (see §2.5), wherein one chooses between, say, civil and chemical engineering. Typically, each segment has its "group," or, in the case of Faculties, its "department," each of which is responsible for teaching that particular segment.

---

[2] It has almost become politically incorrect to use the word "discrimination." This makes writing difficult, because everyone discriminates continuously all day long—between the left and right sides of the road while driving, for instance. It is only some kinds of discrimination—based on race, for example—that should be eschewed.

While this book does not wish to argue at all against the historically pragmatic and eminently practical "discipline" segmentation principle, it does seek to promote another important principle—the one based on what might be called the "engineering role" or {applied science, design, management} segmentation principle, illustrated in Fig 1.6. An engineering school should **not** feel it has to choose between these two principles. Engineers understand vector-matrix thinking and can handle two segmentational dimensions at once. All but the smallest schools should have some response to both of these career (and therefore program) dimensions: engineering **type** and engineering **role**.

Figure 3.1: The 2D Matrix Underlying Ideal Curriculum Choices.

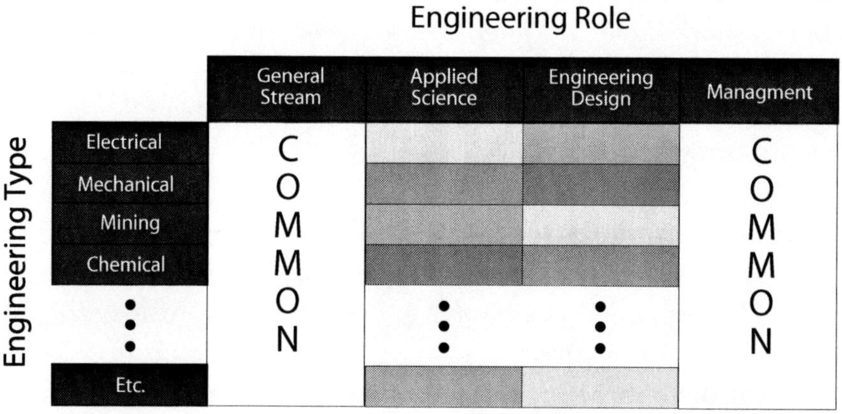

		Engineering Role			
		General Stream	Applied Science	Engineering Design	Managment
Engineering Type	Electrical	C			C
	Mechanical	O			O
	Mining	M			M
	Chemical	M			M
	⋮	O	⋮	⋮	O
		N			N
	Etc.				

The matrix is shown explicitly in Fig 3.1, as would apply in the senior years after introductory years have been completed. The three engineering role pillars are conspicuous above, and a "general stream" has also been added for students who don't want to specialize as to type of engineering. This won't please some staff denizens in individual departments (if they are, as usual, defined by engineering type), but their pleasure should not be the mission of the school. This general-stream idea is a bit off-topic here and won't be given further attention.

Engineering students, who are the customers in this discussion, would benefit greatly from a matrix of program options like the one shown above. They could tune their course and program choices more precisely to their career interests and needs. Ideally, they could choose (as examples) the civil/management emphasis, or the electrical/design stream, or the aerospace/applied-science path. Even though Fig 3.1 is hardly representative of the curriculum in engineering schools, students will benefit, it is hoped, from this discussion by choosing their isolated optional courses (to the extent that such are allowed) in accordance with some of the ideas being discussed here.

As pointed out in §2.5, the "engineering role" segmentation principle is already in place in several successful engineering schools with respect to *applied science* programs, although (to the author's knowledge) there are no schools that have really paid much attention to either the *design* or *management* aspects of this same principle.[3]

Typical Actual Undergraduate Curriculum

With the curriculum structure shown in Fig 3.1 as a target, let us examine a typical undergraduate engineering curriculum for comparison. We shall use the Canadian jurisdiction, with which the author is most familiar, where engineering curricula are policed[4] by the Canadian Engineering Accreditation Board. The CEAB's primary purpose is to identify[5] "Canadian undergraduate engineering programs that meet or exceed educational standards acceptable for professional engineering registration in Canada."

---

[3]This shows, incidentally, a further demonstration of the bias toward applied science because, in principle, a management stream or division should be the easiest to implement, because it has so many elements that are common across all the engineering disciplines (types).

[4]In Canada, a graduate from an engineering curriculum, even at a prestigious university, cannot use that curriculum as part of registration as a *professional engineer* unless the CEAB says it can. This gives the CEAB enormous power and severely limits what engineering schools can do if they wish to retain the blessing of the CEAB. Fortunately, management schools have no CMAB to stifle creative new curriculum ideas.

[5]From "Accreditation Criteria and Procedures," CEAB, 2004.

Figure 3.2: Typical Undergraduate Engineering Curriculum (≥ 1,800 hours).

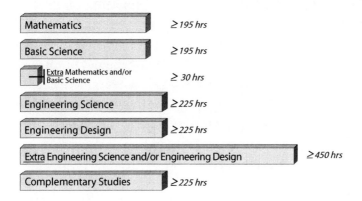

The CEAB places all undergraduate engineering courses into one of five categories: **(a)** mathematics; **(b)** basic science; **(c)** engineering science; **(d)** engineering design; and **(e)** complementary studies. The first four of these categories are roughly equivalent to what we are calling **(a)** abstract mathematics; **(b)** science; **(c)** applied science (including applied mathematics); and **(d)** engineering design.

Thus, the CEAB further refines "applied science" into three subcategories (which it calls "mathematics," "basic science," and "engineering science"), and agrees that design is important to engineers; however, as we shall see, "complementary studies" is a very poor proxy for "management."

The CEAB also places minimum requirements for time spent on each category. The first four categories include two of the areas shown in Fig 3.2—engineering design and applied science (including applied mathematics), which consume half the program time. Also included in the typical curriculum, however, are (basic) science and abstract mathematics as large time consumers (at least one-quarter of total program time).

In contrast, management (not to mention its components) is nowhere to be seen. It does exist, a tiny bit, in principle, in the complementary studies component, typically allocated about 15 percent of the curriculum. This module covers a lot of territory, however. To the CEAB, the phrase "complementary studies" includes

> "... studies in humanities, social sciences, arts, management, engineering economics, and communication. While considerable latitude is provided ... the curriculum must include studies in engineering economics, the impact of technology on society, and subject matter that deals with [the] central issues, methodologies and thought processes of the humanities and social studies. Provision must also be made to develop each student's ability to communicate adequately, both orally and in writing."

Two implications are quite clear from this recipe: first, management is indeed mentioned, though quite feebly; and, second, although the aims behind each of the above long list of possibilities are doubtless meritorious, once the several *required* complementary studies pieces have been satisfied, the time left for anything resembling management studies is not materially distinguishable from zero.

There is, in short, a tendency in engineering schools virtually to require, by the process of exclusion, that undergraduate students not study management at all.

3.2 GRADUATE CAREER CHOICES

After undergraduate engineering, the student either leaves to enter the workforce or continues studying toward a graduate degree. As described in Chapter 2, there is a strong cultural bias towards the view that "success = PhD." In other words, the applied science bias tends to promote the PhD as the logical framework for further study. For many students, this is the right choice, and we all benefit from the creativity and inquisitiveness of those who investigate the application of science to societal needs. For other students, however, those interested in a management flavor to their career—the ones for whom this book is written—a PhD is an odd choice for progress.

MBA vs. PhD

To repeat once again, the discussion at this point relates *not to all* students, but to those who **(a)** are interested in management, and **(b)** would like to pursue graduate study. In the following, "PhD" is used as a proxy for "graduate study in applied science," and "MBA" is used as a proxy for "graduate study in management." Both degrees may require some intermediate experience after the BASc—a master's degree in the case of the PhD, and some work experience in the case of the MBA—before entrance is permitted to the desired advanced program.

While serving as the founding director of the Jeffrey Skoll BASc/MBA Program at the University of Toronto, the author had occasion to discuss this issue, one on one, with hundreds of students, each of whom was interested in (or thought they *might* be interested in) management as a career path. All had either strong or outstanding academic performance, and virtually all had a strong résumé in outside (i.e., nonacademic) activities.[6] One question that arose with surprising frequency was this: I would like to get a PhD and an MBA; which should I get first?

The answer to this question evolved into something like the following:

> You have enormous energy and vast potential, and I understand why both these degrees are of interest to you. You are accustomed to excelling in all you do; it is natural therefore that, since you are excellent in your engineering [applied science] studies and also have good reason for thinking you are interested in and would be successful at management, you are interested in both a PhD and an MBA.

---

[6]These students tended to excel at "outside activities." These included athletics (at the varsity or national level); high achievement in the arts; significant leadership in student, political, or other organizations; and community involvement (including charitable work). It was difficult to escape the conclusion that, at least for this group of engineering students, there was a very high correlation between interest in management and general leadership interests and abilities. Schoolwork was a sizable focus, but only one of many.

Chapter 3: Important Decisions in Engineering School

You are talented enough that you would certainly have a fine career even without further university study, but I will assume that you do indeed go on to graduate work at some point in the near future.

I know I don't have to tell you, a smart engineering student, how to calculate the many years and the many dollars your idea of getting both a PhD and an MBA will cost you. Don't forget that a PhD will take a large fraction of a decade, and since you have only about 45 years of working life at this point, that's a huge commitment, especially if an MBA is also in your plans. As for costs, you have to include the opportunity costs, meaning in particular all the income foregone because you are in school. And I know you don't want to go through all this hard work and end up working for someone who skipped all this schooling and went straight to climbing the career ladder in the workforce. Especially if you think you're smarter than he is!

Figure 3.3: Breadth vs. Depth: [left] an archeological dig; [right] Australia, compiled from 165 RadarSat satellite images (2001).

But here's a way of looking at this issue that you may not have thought of before. In one important respect, a PhD and an MBA are the opposites of one other—breadth vs. depth (Fig 3.3). As an analogy, think about an archeological dig, and consider the enormous attention to detail it requires. The digging, the brushing, the washing, all painstakingly necessary to accomplish the research goal of high-quality peer-reviewed archive-publishable research. By comparison, think about satellite imaging from space, and compare the scale of this information from space with the size scale of the "dig." Geometrically—and that's the analogy—one area of study has a scale, or breadth, four orders of magnitude greater than the other.

It is not helpful to ask, "Which is better, archeology or satellite imaging?" That's like asking, "Which is better, the piano or the violin?" Ultimately one has to choose one's field, and the most helpful questions to ask oneself are, "Which instrument do I enjoy more?" and "Which instrument do I play better?"

Figure 3.4: Breadth vs. Depth: PhD vs. MBA.

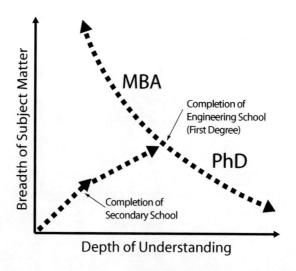

That's why I feel it is rather inconsistent to get both a PhD and an MBA. They are almost opposites of one another in their approach to learning and career development (Fig 3.4). My advice is that you really should choose one of them and leave some career time to get the full advantage of that *one*. If you get a PhD you can always work for an MBA grad, and if you get an MBA you can always hire a PhD grad to do the research.

Many of these students gave the impression that they were potential managers trapped in an applied scientist's body—or at least in an applied science program. The perception was of some cognitive dissonance between their natural impulses and the culture of the engineering school.

Immediate Employment vs. PhD
The holy grail of most modern engineering schools is the PhD, and this is the correct choice for those who have the drive (and the talent) to undertake a career in applied science (a.k.a. engineering research). The only warning needed, perhaps, is that a dram of caution should be exercised with respect to an organization whose first degree in "engineering" is the Bachelor of Applied Science and whose highest earned degree in "engineering" is a doctorate in philosophy. Not all engineering schools use this terminology; one can almost hear the echoes of faculty arguments when MIT decided to award the DEng degree, saying, in effect, that a doctorate in engineering should be just that.

The PhD in engineering is unashamedly an applied science degree, which is fine as long as students are told this very clearly and are also made aware of other promising program alternatives. An engineering PhD thesis, like engineering research papers, is judged by the *science paradigm*, and its character is almost indistinguishable from a PhD thesis in science or applied mathematics. Theses that lack an avalanche of matrices, or partial differential equations, or something suitably Boolean or Bayesian (not to mention Besselian), are deemed to be seriously deficient, whether or not they contain creative designs, or solutions to practical engineering problems—or, in fact, regardless of any other positive attributes possessed by either the thesis or its putative writer.

Winners in this sweepstakes—the holders of new engineering PhDs—find that they have now finally come to the end of the graduate-study road and that they must, as did those undergraduate classmates from whom they parted long ago, find a place of gainful employment. The idyllic place to situate is, of course, a faculty position at an engineering school.[7] Some are successful in this quest; these are the young PhD engineers who have survived the gauntlet of proving that they have the mathematics, science, and research attributes valued by the academic-engineering culture. This is all as it should be, except for one important thing: these applied science researchers should be only *an important part of*, not the exclusive denizens of, an *engineering* school. A PhD in engineering indicates a proficiency in applied science research. Not in engineering. Not in teaching. And certainly not in management.

A PhD is now mandatory for a teaching/research position in a university, so if that's where one is headed, the choice of a PhD must be made. Some PhD graduates find employment in companies that are interested in their research area, although most fresh PhD's are so narrow that a major paradigm shift is required to make them really fit in. Others will work in government labs. Still others just below the cutoff line of success become situated in post-doctoral positions, an antechamber from which some later emerge into permanent positions after several years of additional experience—and others do not.

To make the most important point one more time: There is nothing wrong with a PhD in engineering; it can lead to a gratifying and rewarding career for those with the right talents and interests. The caveat urged upon the undergraduate student here is this: Know what a PhD is and what it isn't; carefully consider the costs and the rewards; evaluate honestly your abili-

---

[7]Despite all their training in, and loyalty to, mathematics and science, virtually no one who emanates from this process ever seems to take up a position in a mathematics or science department.

ties and aptitudes; don't be taken in by the easy assumptions of others whose interests may not be quite the same as yours. You are choosing a half-century of your life. If, after all this introspection, you remain convinced that a PhD in applied science is right for you, go for it! You will always know it was a carefully considered decision.

3.3 IT'S NOT ROCKET SCIENCE—IT'S MORE DIFFICULT!

The prevailing culture in engineering schools tends to favor mathematics and science abilities and achievements. This has been stated several times in the foregoing, and is eminently reasonable for those engineers who choose the applied science path. Maths and science dexterity at a truly extraordinary level are unarguably needed to produce applied science research that is of world class. One cultural value in the typical engineering school that seems to accompany that fact: Maths and science are thought very difficult, and all other academic pursuits are thought rather straightforward. Into this latter category fall management and its constituent subject areas.

Mathematics vis-à-vis Management

Certainly mathematics seems, like music, to have a hierarchy of excellence that is more elaborate and extensive than most other intellectual pursuits. No matter how good one is at mathematics, there is always someone else who is much, much better (unless your name happens to be Archimedes, or Newton, or Gauss, of course). It would be interesting to compare mathematics to management in these respects, but unfortunately this is quite difficult. This is due to two basic differences between mathematics and management.

First, the purview of mathematics is much narrower than senior management. Yes, mathematics has been defined as "the study of patterns," which sounds pretty broad, but there is almost no area of human interest that is completely unrelated to management. There are also aspects of management talent that are more inchoate, less subject to classification. As one example, consider *charisma*. This is one of those words that is difficult to define, but as Sir Kenneth Clark remarked about civilization in his book *Civilisation*: it is difficult to define it but you know it when you see it.[8] Charisma is helpful to a leader or manager. Yet it cannot be taught—in engineering school or anywhere else.

Second, mathematics is a totally rational subject—surely its greatest attraction—and this implies that one can, for example, distinguish between a correct proof and an incorrect one. A correct proof gets complete respect, as it should, while an incorrect proof gets zero marks (except in school). Words, like "correct," "incorrect," or "perfect" rarely get used in describing management processes or decisions. For reasons explained in more detail below, most management happens in a metaphorical fog within a partial vacuum: a great manager is one who is right more often that he is wrong.

Which Is The Bird Course?

With these preliminaries now on board, let's take a look at "management" vis-à-vis "applied science" with attention to how they compare as course offerings in post-secondary school. The assumption among many engineering students, who daily survive the struggle with calculus and the war with physics, is that management-related courses (and anything else, actually, that is not applied science) are probably *bird courses*, meaning, one supposes, something fragile and flighty, and certainly not the sort of thing a serious engineering student needs or wants.[9]

Consider Fig 3.5. Which is the "bird" course?

---

[8]*Civilisation: A Personal View*, Harper and Row, 1969. To be more accurate, what Clarke said was this: "It may be difficult to define civilisation, but it isn't so difficult to recognize barbarism." (p 241)
[9]Oddly, they usually get higher grades on the math courses than on the allegedly "bird" courses. This counterintuitive result is rarely contemplated, much less satisfactorily explained.

Chapter 3: Important Decisions in Engineering School

Figure 3.5: Which Exam Is More Difficult?

	Typical Math Problem	Typical Management Problem
Enough Data?	Always	Never
Too Much Data?	Never	Sometimes
Reliable Data?	Always	Rarely
Up-to-Date Data?	Always	Rarely
Correct Answers?	Always	Rarely
Precise Concepts?	Always	Fuzzy
Issues Few?	Always	Rarely
Issues Simple?	Always	Rarely
Formulas Available?	Usually	Rarely
Many Specialities Involved?	Usually 1	Usually $\gg$ 1

If math exams were prepared similar to management exams, one might be faced with quiz problems like the following:

> Find the area of the following geometrical figure. The figure, to be simplistic, is a circle, but is probably more like an ellipse—and a multi-dimensional ellipsoid may be what we're really talking about here. To be honest, this whole ellipsoidal business is just a geometrical approximation. Who knows whether this shape is really that simple—or even roundish, for that matter?
>
> If you want to get any actual numbers though, think circle with $r = 5$, according to Fred, the CEO. On the other hand, Lisa claims that she has calculated $r = 4.7$, and she should know because she's an engineer. Sam, the accountant, says that $r = 9.13847$—a good example of the difference between accuracy and precision! And Sally, our expensive consultant, gave us $r = 3$ on Wednesday, and $r = 4$ a week ago Friday. Who knows? Those were the data we had last month; they likely are out of date by now. (With information like this, the idea that $\pi = 3$ has more accuracy that you need.) Anyhow, look at the above numbers, and do your best. In the end, you may have to consult your *gut*. Call on all your experience.
>
> Here's the grading scheme. If you get this wrong, the company's market value will take a nosedive. Your shareholders who are pensioners will suffer. Your suppliers may start to demand cash. Your

customers will wonder whether you will be in business next year. And your employees will be perceived as too numerous.

Probably the R&D department will be the first to go. What's the use of a long-term investment like that? If you don't get pretty close to the right answer on this question, there may not even be a medium term.

So, which is the bird course?

Side Benefits To Studying Management
Even for engineering students who do not plan to climb the infamous corporate management ladder, there are still many benefits to taking a few management courses. These include the following:

1. If you plan to work in a business—be it a small firm or a giant corporation—it is in your interests to know as much as you can about how the organization functions. If you were going to spend forty hours of every week of your working life on a sailing yacht, would you have no curiosity about how it functioned? Even if you planned never to become the skipper, would you not be interested in fundamental facts, such as that you could never sail closer than 42° to the wind, or that you had to keep a sharp eye on the boom going downwind in a variable Force 3 wind because an unexpected jibe could bring the boom around suddenly and knock you unconscious or out of the boat? Knowing how businesses function generally will prepare you to learn how the one that employs you functions.

2. Many non-businesses are really quasi-businesses. For example, suppose you work for a not-for-profit organization. It still has a budget and financial responsibilities. It still has governance, a management structure, and human resources functions. It still must market its product or service in some fashion if it wishes

not to shrivel. And, unless it is in a powerful monopoly position (the government comes to mind), it must operate with some degree of efficiency. Consider a hospital: true, it may have, as it primary objective, caring for the seriously unwell (rather than making a profit for its shareholders, as with a private for-profit corporation), but running a complex operation to meet a hospital's complex objectives and earn zero profit is not easier than running a complex operation to meet the simple objective of making 8% profit on revenue. Or, consider a church, for which one frequent sales tool is the promise of a life after death, which never ends, in a very nice place. Though these enterprises may be not-for-profit, they have quite sophisticated marketing strategies. The Roman Catholic Church has its own bank. Altogether, management principles apply as much to not-for-profits as they do to for-profits. So, even if you are headed for one of the not-for-profit sectors, you should still be interested in business and management.

3. If the organization that employs you is well run, it will value you (and reward you) to the degree that you add value to that organization.[10] Most (but, strangely, not all) employees want to be highly valued and well rewarded. (Others would like to do just what they feel like doing.) It follows that you should have a mental model of your organization, of what it considers valuable, and of how you can contribute to that value. This mental model will not come from studying Bessel functions (to use our favorite whipping-boy example). Here is an interesting fact about Bessel functions:

$$J_n(z) = \frac{1}{\pi}\int_0^\pi \cos(n\theta - z\sin\theta)d\theta$$

where z is a complex number and n is a natural number. This is an elegant formula and irresistible to anyone with mathemati-

---

[10] Unless you are the son or the brother of the owner.

cal interests. The question at issue, however, is this: should any engineering student ever spend study time on this and similar material (unless doing a doctoral applied-science thesis based on cylindrical geometry)? Yet many students find themselves involved with endless obscure items like the above rather than the sort of material that they would find very useful, *continually*, in their careers. The above formula does not explicate how to add value to one's organization. It does not give insight into the impact of one's technical (or other) work.

4. Another helpful thought experiment might be this: Picture all the front pages of all the (serious) newspapers you will ever read. How many of these do you think will feature at least one article for which an understanding of economics would be a great asset? (Probably most of them.) Then why would you not be almost desperate to acquire at least some understanding of microeconomics and macroeconomics? Shouldn't you use your university education to help you to understand the world you will be living in for the rest of your life?

5. Finally, if you harbor any wish to start your own enterprise at some future time in your life (see Chapter 12) then you should be starving for management courses. To be a successful entrepreneur, you will have to learn truckloads of information and concepts, for the rest of your life. (Fig 3.4 may be helpful here again.)

3.4 A CONSTELLATION OF PROFESSIONS

We have seen in Chapter 2 that the word *engineer* often means different things to different people. To engineering professors, it often means someone who can show how to apply scientific principles to create exciting new possibilities for solving human problems and needs—in short, someone like themselves. To engineering designers, it would likely mean someone who can combine the art, science, craft, and efficiency of engineering design to create a result that meets agreed specifications, that can be manufactured with high quality but low cost, and that has ready customers or beneficiaries for their organization—in short, someone like themselves. To engineering managers, it might mean someone who can lead a team to create the prototype of a complex new product or service, be it a new aircraft, a new computer, or a new brewery—and despite the multitude of rules and regulations, uncooperative groups, distractions, sundry naysayers, and diverse threats—on schedule, on budget, and within specifications. In short, someone like themselves.

Principle Of Vocophilia

This pattern suggests an important principle: the genus *engineer* comprises a *dramatis personae* of closely related characters whose roles are viewed slightly differently by each member of the group. (This general observation was exemplified in a particular case earlier, in Fig 2.1.) These variants on what it means to be a "real" engineer should be interesting enough to any engineering student who is not always staring at his own shoes, but the *Principle of Vocophilia*,[11] as we shall call it, can be taken much further:

> **Principle of Vocophilia**
> Every profession views itself as being somewhat **more difficult to learn**; somewhat **more difficult to practice**; and, generally, somewhat **more important to society** than other professions (and the general population) view it to be.

This principle is the result of the observation that, almost without exception, each profession sees itself as more prominent in society than its sister professions see it. Generally, each profession has a

---

[11]Although apparently a neologism of the author's, *vocophilia* represents an important idea where no other word seems to exist. It is similar to *topophilia*, meaning "We love to stay where we have lived for a long time."

view of itself that is a somewhat amplified version of the view seen by society at large, including other professions.

Figure 3.6: Bar Chart Illustrating the Vocophilic Principle.

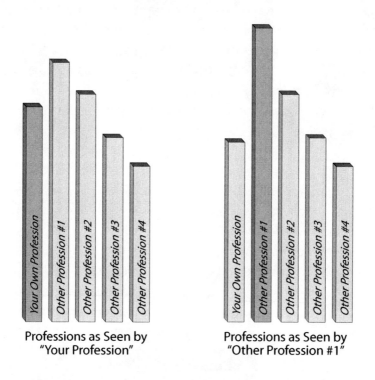

Professions as Seen by "Your Profession"

Professions as Seen by "Other Profession #1"

This is illustrated in Fig 3.6. The point here is not to blame anyone, or demean any profession. We are simply recognizing human nature. Everyone knows the challenges and contributory role of his or her own profession better than members of other professions could be expected to.

The intention here, instead, is to point out to engineering students not only the parochial culture and values of the typical engineering department—whose provincial shortsightedness was highlighted in the last chapter—but also to alert students to the many other categories of professional persons they will frequently have occasion to contact in their working lives. Students should welcome such contacts if they plan on developing as competent organizational team members, and if they wish their professional life to be interesting. One practical way to acquire some early experience of this type is with the aid of the apprenticeships available from temporary co-op positions.

Other Key Professions
From a management perspective, many professions are of daily significance (in addition to the several flavors of engineers). Accounting is of such importance that Chapter 8 will be devoted to it later, in Part II, and there are several smatterings from the law profession as well (particularly in Chapters 10–13). How do these professions see the world (meaning the business world) and how are these perspectives different from those of most engineers?

One thing can be said of the professions of accountancy and law—and, for that matter, of all other professions: None of them should, by virtue of the profession alone, have the hammer in all general business decisions. There are many examples of instances in which a recommendation from one of these professions, however excellent it may be as judged by the parameters of that profession, is not the correct decision when all the relevant business data are considered; there are many other examples where their advice is ignored at one's peril.

It takes courage to overrule the opinion of a competent lawyer in a business decision, and it takes nerve to ignore the advice of an accountant in a commercial situation, but there is only one profession[12] entrusted to judge all the inputs from these and other professions and that profession is "senior manager," meaning "officer of the corporation." In fact, if the decision is sufficiently large—meaning for

---

[12]On the other hand, *renegade* senior managers—as typified in the now-familiar WorldCom and Enron debacles—who ignore everyone and everything, especially their consciences, in the pursuit of unprecedented levels of personal greed, is a quite different phenomenon. These are not outstanding senior managers who balance various professional opinions to arrive at a difficult but brilliant decision; these are just white-collar crooks who exploit their positions of great trust to steal millions of dollars from their employees and shareholders with the sole intent of enriching themselves personally.

enough money, or an important policy matter—the decision is too large for any one person; it must be made by the Board of Directors.

3.5 TECHNOLOGY AS VIEWED BY OTHER PROFESSIONS

To provide some insight into the differing viewpoints of various professions, we now take a short and somewhat light-hearted tour of the word *technology*. If engineers, technologists, engineering professors, accountants, lawyers, finance experts, and venture capitalists were each asked to do a word association test with respect to the word "technology," what might the results be? Or, to use a more practical scenario, if one were at a conference table with members of the above-listed professions to consider the advisability of pursuing a new technology, what sorts of issues would each profession raise? Where would each profession likely be most helpful?

Once one has some conception of these instinctual differences, one can better read the comments of each of one's non-engineering colleagues. One can tell "where they are coming from," at least professionally.

Engineers And Technologists

Although engineers and technologists have, even among themselves, different emphases, interests and insights, both might have the same response to Fig 3.7. Which is the better technology?

Figure 3.7: How Might an Engineer View a New Technology?

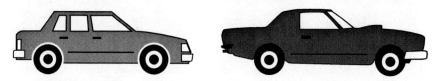

The car on the left has relatively modest pretensions, and is sold at a price common customers can afford. The fine automobile on the right is clearly aimed at a much higher level.

Chapter 3: Important Decisions in Engineering School

Technologists drool and engineers swoon at the technology represented by the right-hand machine because its technology *pushes the envelope*. Its top speed and 0–60 acceleration set new standards; it has gadgets and toys not previously seen outside the racing-car circuit; and it will convince colleagues and possibly some members of the opposite sex that its new owner is, by association, attractively special.

Technically trained persons are, understandably, interested in and impressed by technical performance characteristics. Yet the modest car on the left may sell 1,000 times as many units as the automobile on the left and may therefore be a better business product.

Engineering Professors
Lest we forget that this Part I is written primarily for engineering students—and that the most influential professions at this stage of these readers' careers are therefore likely to be their teachers and research supervisors—we also pause and ask through what means these professions might view the significance of a new technology. A likely answer, one that would be ridiculed in public but that is followed rigorously in private, is suggested by Fig 3.8.

Figure 3.8: How Might an Engineering Professor View a New Technology?

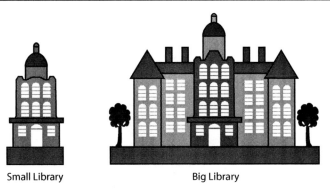

Small Library　　　　　　　Big Library

Engineering professors are judged based primarily on the prolificacy of their research publications, as are the rest of the applied science research community. Not to over-simplify, but the basic process is counting papers. Reliance is laid on the principle that the more difficult the new technology, the more research was needed (as measured by the number of research papers). A whole book could be written (and probably should be) on the pros and cons of this approach, but we shall be content here to observe **(a)** that "difficult to develop" may be a characteristic only tenuously correlated with ultimate utility, and **(b)** research papers are published based on the recommendations of reviewers who are themselves also applied science researchers. This *peer review* process is (as was said of democracy) the worst possible process except for all the alternatives. Inbreeding among the reviewer class is not just a danger, it is a fully accomplished fact—both as to specific topics favored, and, even worse, as to the broad values employed.

Accountants

We shall have occasion later (in Chapter 8) to look more closely at accountancy and some of its roles in management, but for now we ask merely this: How might an accountant value a new technology? Fig 3.9 symbolizes the answer. The answer, in brief, is to add up all the dollars spent on its development. These dollars will have been painstakingly counted, because counting is the objective part of accounting, and it is easy to see how this has historically come to be.

Figure 3.9: How Might an Accountant Value a New Technology?

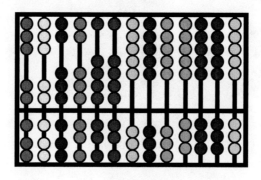

Chapter 3: Important Decisions in Engineering School

When dollars are involved, human beings are at their most creative. The most fundamental goal of the accounting profession is to prevent one person from stealing, or otherwise purloining through more sophisticated schemes, another person's cash. This means keeping one's eyes not only on the dollars, but on the pennies as well, leading to precisions (to the penny) that seem rather humorous to someone with mathematical training.

The numerical values used are also *transaction-based*; two independent (arm's length) parties have agreed to the value of the transaction. This meets the burden of the *principle of objectivity*, which is very important in accounting. This principle is implemented here because **(a)** there is no outside bias, **(b)** the result is credible, and **(c)** the dollar value associated with the transaction is immutable (will not change over time). Obviously these are important concepts, and we can, without overstressing our imaginations, make out how these values came to govern the profession of accounting.

However, the answer to the original question (What is the value of this new technology?) tends to be something along the lines of "$4,248,576.32," based on all costs incurred in its creation to this point. This contains some useful information, but is hardly the final answer to the question, from either a management or a societal viewpoint. Whatever it cost to develop, if this technology does not lead to commercially viable products, its value is $0.

Lawyers

Once again, we shall later examine in more detail (e.g., Chapters 10–13) some of the relationships of law to management, but here we just ask this: How does a lawyer go about the process of technology valuation? Fig 3.10 contains a prosaic depiction of at least part of the answer.

Figure 3.10: How Might a Lawyer View a New Technology?

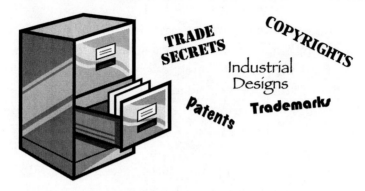

Lawyers rarely get in the game of evaluating technology. They leave dollar values to accountants and others. What typically concerns them, instead, is this: "Who owns it?" And, "How can we ensure that no one else steals it?" It is hard to deny that these are important related questions.

The law profession has greatly enhanced the notion of "ownership" over the millennia. For example, the Ten Commandments of western culture (invented originally for men only) speak of a man's possessions as including sons, daughters, manservants, maidservants, and cattle. One is also prohibited by the Decalog from coveting one's neighbor's possessions, and the explicitly mentioned items on the list include: his house, his wife, his manservant, his maidservant, and his ox. This is an early balance sheet (assets only) of what a man needs to carry on a reasonably good life. The law has evolved by orders of magnitude, in both focus and complexity, over the past millennia,[13] but the basic idea that one of the chief functions of law is to identify items owned and to devise documents and procedures to protect them from thievery has remained unchanged.

One of many parts of modern law of interest to business and management is concerned with a category that would never have

---

[13]For example, wives are no longer owned by their husbands in most advanced cultures.

occurred to any but the most brilliant person of antiquity: ownership of items of the type we now call *intellectual property*. Thus, when asked "What is the value of this new technology?" a lawyer is likely to respond, "I do understand that it is of considerable value, but what I would rather help you with is this: How do you know that this new technology will actually be yours next year?" In other words, "I can help you guarantee that you and your organization will be the ones who will benefit financially from it." We shall have a closer look at intellectual property in Chapter 10.

Business Valuators And Investors
Among those who have a great interest in the value of a new technology are all those who place a significant financial wager on its business success. This group includes stockholders (a.k.a. shareholders), angel investors, venture capitalists, the public markets, pension funds, and government agencies charged with stimulating business growth and productivity improvements.

Chapter 9 will discuss the process of technological *innovation* in more detail, but for now we shall be content to observe that, to those who put actual cash into a new business, it is actual cash out in the future (and more out than put in) that is the ultimate measure of the success of the new technology.

Figure 3.11: How Do Investors Value a New Technology?

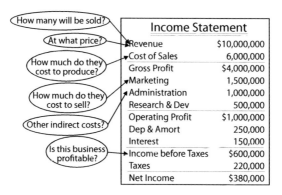

Thus, for investors, the value calculation follows an algorithm along the lines of the one shown in Fig 3.11. To create wealth, a new technology must result in products (in the most general sense) that sell in a free market for more than it costs to produce them.[14] Some may scoff at this so-called bottom-line approach, but the scoffers never seem to be the investors, so perhaps there is a free lesson there. Sometimes the scoffers want the government to be an investor; in other words, the scoffer advocates that *other* people's money be invested in the new technology (likely for some reason that favors the scoffer). As it turns out, sometimes money contributed by the general citizenry *should* be invested in a new technology, but only when the same basic criterion that every other investor uses to measure success is used: the so-called bottom line.

3.6 IN SEARCH OF A BUSINESS MODEL

Engineering students learn many skills that stand them in good stead, not only within their own field of study, but more generally in life's many situations. One of these skills is the concept of a *quantitative model*, wherein a process of interest is described using *inputs* and *outputs*. The workhorse of this approach to modeling processes is the *block diagram*.

In Praise Of Block Diagrams

A block diagram breaks down a process of interest into several sub-processes with the inputs and outputs identified for each sub-process. The most elemental unit of a block diagram is a single block, with its inputs and outputs defined. In practice, the outputs are of two varieties: *measurable outputs* (what most would more economically be called *measurements*), and *controlled outputs*, i.e., what we are trying to control. Unfortunately, we cannot always measure exactly

---

[14]We shall further explain the idea behind these "profit or loss" calculations—referred to formally as "income statements"—in Chapter 8.

what we would like to, and this is the basis for the distinction between what we can actually measure or sense (the measurements) and what we feel is the object of the game (the controlled outputs).

In order to make the model completely quantitative, the details of what goes on within each block must be specified quantitatively (i.e., mathematically) and the number of possibilities here are numerous: amplification factors; simple algebraic relationships; time delays; differential equations ("ordinary" or "partial"); difference equations; lookup tables with interpolations schemes. At the other, less-quantitative extreme, and not an infrequent situation, we just have an expert sitting at the table who knows more than anyone else does about what is (or may be) going on. In general, the author will assume that all readers of this Part I have a sophisticated impression of the uses and abuses of block diagrams.

Is There A Generic Business Model?
The next question is naturally this: Is there a standard block diagram that can model a generic business? If so, are there well-known sub-blocks for the processes and effects that management in general (and engineering managers in particular) are supposed to bring about? The short answer is "No," and in fact, no one (to the author's knowledge) is working on this problem. People in business simply do not think in these terms.

True, there are places in business where this sort of analytical thought has proven useful, and where substantial developments have occurred. (Obviously, there are places in R&D where math models are helpful, but this is hardly what we're talking about here; we are asking whether there are math models of the business *itself*.) *Operations management* is one area where mathematical principles are of practical significance, and, indeed, some of the mathematical difficulties associated with queuing theory, for example, have attracted first-rate applied mathematicians. In addition, no discussion of the application of mathematical modeling to business can proceed very far without the advances in the *finance* area being cited.[15] If one examines the Nobel Prizes for Economics for the past few decades, one is amazed by the mathematical simplicity of some of the results that led to this great honor.[16]

Basic Financial Business Model

The subject of the mathematical modeling of a generic business (or even of a particular business) will be left aside here, owing to a lack of serious material, but we shall exit with at least one block diagram model—the overall financial model of a business. Even at this simple level, most engineering students will benefit from reflecting on its significance. Only major financial inputs and outputs are shown.

This model is shown in Fig 3.12; horizontally are shown the inputs of direct interest to an investor (a.k.a. stockholder or shareholder). The input investment, made at $t = t_1$, is expected to be returned at a later time, $t = t_2$ (thus, $t_2 > t_1$). In fact, at the later time, the investor expects to get more out than was invested. This extra amount is called the *return on investment*, or RoI, and the reasonable value of this RoI depends on the waiting time $(t_2 - t_1)$ and on the level of risk. This aspect will be discussed in Chapter 7.

Figure 3.12: Simplest Model for a Business.

---

[15]It is noteworthy in passing that there are many happy examples of engineering students who decided to study management, and who ended up in exciting careers in either finance or operations research.
[16]For example, what is now referred to as the Black-Scholes model for the valuation of financial options has the mathematical complexity of a take-home exam for a senior undergraduate honors physics or engineering science student. Yet it won the Nobel Prize in economics in 1997.

Vertically in Fig 3.12 are shown the principal operational cash-flow categories. Sales to customers and loans from lenders (banks, etc.) represent cash into the company, while payments to suppliers and salaries and wages to employees represent cash-flow out of the company. Also shown are interest payments to lenders and corporate income taxes[17] on profits earned; from a business standpoint, these are "leaks," although not thought of thus by their recipients.

A Call To Arms
Ultimately, the success of the business modeled in Fig 3.12 depends on the internal "miracles" that occur. Think of Fig 3.12 as a *game module*. The Miracles—meaning processes that add net value, as measured by the customer, and thus create wealth—are all the work of human beings, aided by technology and guided by management. They are very difficult to explain (otherwise, computer stores would be selling "business machines" for making money).

The business shown in Fig 3.12 has competitors in its own business sector; they each also have a similar game module. Using the substitute principle of microeconomics—which says that the demand for Good B will increase if the price for Good A is increased—we see that many other businesses are also involved in the economic climate of any one business. Finally, in this modern era of globalization (a centrally important process that no amount of wishful thinking can prevent from happening), the business in Fig 3.12 is in competition, directly or indirectly, with virtually every other business on the planet. Protectionism, socialism, and other varieties of navel-gazing and wishful thinking are not credible solutions. *Increasing productivity through exploiting technology* is what's needed. In this campaign, engineers should naturally be vital participants. Not just as engineers in the narrow sense, but as managers and leaders, at all levels.

---

[17]Other taxes, of which there are many, are absorbed into other prices and cash-flows in this model.

CHAPTER 4

The Long Path

From Eureka to Ka-Ching

It would be asking much too much of students who have just freshly enrolled in engineering school to realize the myriad interrelationships between engineering and the other professions, occupations and vocations with which engineering ultimately interacts. It would be asking even more of them to consider how this interaction should best take place—both generally, and *for them* in particular. Finally, precocious dimensions of intellect would be needed were they asked to understand all the intellectual connections between what they will be studying in engineering school and the massive accomplishments, in related fields, upon which their studies and career rest: millennia of mathematics; centuries of science; and decades of innovation in engineering and technology.

Still, they *must start* at the task of slowly comprehending the grand historical context of their life's work. Not just as a satisfying intellectual exercise—although that should perhaps be reason enough as part of a university education—but also, a *fortiori*, with the goal of understanding where they and their careers fit within the larger picture. They cannot make career choices—meaning, at this stage, program and course choices—without this perspective.

Chapter Overview

This chapter will be an elaboration of Fig 4.1; we shall describe the Long Path from historical mathematics and science to modern technology and commercialization. Perhaps the term "concept chain" would also be an apt metaphor for how ideas get passed up through the diagram by a multitude of people and professions over many, many years.

We shall first spend some time looking at the foundations upon which modern engineering rests. Engineering school curricula typically do a pretty good job in this area already, as would be expected with their applied science bias (Chapter 2). What is usually given shorter shrift, due to the same bias, is an honest revelation to students of the higher levels in the Long Path, and of how other important activities and professions rely on engineering and technology for their continued sustenance.

Figure 4.1: The Long Path from Maths and Science to Commercial Success.

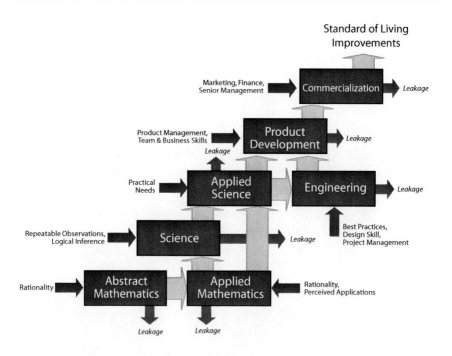

With the importance being placed on Fig 4.1 in this chapter, it is necessary also to stress two further clarifications: **(a)** no one figure and no one chapter could ever capture the richness and complexity of this subject matter; and **(b)** the sole purpose of this graphic model in this book is to connect, at least to some degree, the contributions and roles of the many other individuals and their professions and vocations toward the ultimate goal of the successful commercialization of new technology.

Other viewpoints might well apply a different organizing principle, but we shall be content with the challenges of Fig 4.1. We shall also use this same Long Path approach several times in later chapters. The epiphany in *mathematics*, as represented by "eureka," and as made famous by Archimedes, is the joy of discovery or proof. The apotheosis in *business* is when a customer completes the purchase transaction, as typified by the "ka-ching" of the old cash registers (a sound the youngest readers may never have heard).

Words Do Matter

Fussing with the meanings of words may seem rather boring to some, but perhaps not as boring as listening to a half-hour argument that could be amicably resolved in seconds once the meanings of critical words (and the concepts behind them) were dissected and agreed to. Many consider semantics to be an irredeemably dry subject, of no real interest to practical, action-oriented individuals. On the other hand, it is difficult to have a rational discussion about an important topic when some of the most important words have meanings that drift widely from context to context, from statement to statement, and from person to person.

> Example: When engineering academics use the phrase "research and development" they tend to mean the creation of new ideas in applied science and [if pressed] some potential practical applications for their work. When business leaders use the phrase "research

and development" they often mean market research and product development. There is an enormous difference between these two interpretations of R&D. Conversations about R&D between engineering academics and business leaders can be either highly amusing or highly confusing, depending on one's familiarity with their dialects.

We shall lean towards erring on the side of explaining what these fundamental words mean for us in this book.

4.1 ABSTRACT MATHEMATICS

The oldest of the foundations underlying modern commercial activity is mathematics. It is difficult to visualize any homo sapiens who did not at least count to ten. If humans had twelve fingers, the number basis (and the modern metric system) would undoubtedly have been 12, not 10. We would now have a duodecimal system, not a decimal system.

It is also very difficult to achieve a complete definition for mathematics. Even mathematicians seem not to be in total agreement. Fortunately, our purposes here are served by employing relatively simple definitions that will suffice for our discussion of mathematics, science, applied science, technology, engineering, and management.

Abstract mathematics normally is seen by engineers and most scientists as embracing three sub-groupings, namely: **(a)** arithmetic, **(b)** geometry (including trigonometry and conic sections), and **(c)** analysis in which letters are used to represent abstract quantities (including algebra, analytical geometry and calculus). More generally, we may use the definition shown on the next page. In any case, this first step in the Long Path is shown in Fig 4.2.

The "leakage" in Fig 4.2 refers to mathematical effort and results that are not helpful, either to science, or to any later phase in the Long Path. It is likely that most modern mathematical efforts today, especially in abstract mathematics, are just such leakage—that is, they have no real scientific or commercial or economic benefit.

(Those whose salaries are paid from this activity would, of course, claim otherwise.)

> **Abstract (or "Pure") Mathematics [Definition]:**
> The treatment of the exact relationships between quantities or magnitudes and of the methods by which sought quantities are deducible from known quantities.

Many refer to mathematics as a science; some even call it the "queen of the sciences," whatever that may mean. The author's opinion is that mathematics and science, though often found in cohabitation, are fundamentally different. Mathematics can in principle lead to exact results, unless actual mistakes are made; science is never exact.[1] Mathematics is a collection of pure (and occasionally useful) abstractions; science is always tentative.

Figure 4.2: The First Step in the Long Path—Abstract Mathematics.

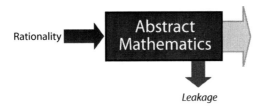

The truth of a *scientific* claim must always be thought of as a *degree of likelihood*, and all claims are subject to further refinement and revision in the light of new data or more sophisticated rational

---

[1]For brevity, this discussion is simplistic. Unlike science, whose object is to elicit truth about the physical, natural world—always approximate, always tentative, but far better than any other approach—the most abstract of mathematics (what might even be called "pure" mathematics) is essentially *metamathematics*. These are pure formalisms (like the rules of chess) and even fundamental ideas like the arithmetic numbers, plane geometry, etc., are, to these formalisms, "applications." Gödel's famous First Incompleteness Theorem proves that no such purely mathematical formalism can produce, by itself, the results of arithmetic, geometry, etc., but in fact require some extra "axioms" that seem intuitively correct but that are outside the formalism. See, for example, R Goldstein, *Incompleteness*, Atlas (Norton) 2005.

arguments. For example, the sciences of cosmology and celestial mechanics suggest that we can rely on the sun rising tomorrow morning with a probability that is less than unity (certainty) by, say, one part in a trillion. In contrast, the area of a circle with radius r is πr^2. This is exactly true; this will always be true; and this will always be exactly true. Nothing about the physical world can be claimed so strongly. On the other hand, there are no perfect circles in nature, so in applying mathematics to science or engineering one must remain vigilant about whether one has an exact solution to a weak math model, or an approximate solution to a better math model.

We shall refer to "abstract" mathematics, rather than "pure" mathematics. (The latter is an excellent example of the Vocophilic Principle depicted in Fig 3.6.) If any purity can be imputed, it is the characteristic just mentioned above—all results are exact. Nasty approximations are not permitted, even if the error is infinitesimally minute—and even if a tiny approximation would permit numbers (quantitative results) to be computed. Once only exact answers are permitted, this severely limits the chances that the problem being solved has practical utility. In fact, some dictionaries define pure mathematics as "the branches of mathematics that study and develop the principles of mathematics for their own sake rather than for their immediate usefulness."

Example 1: The Number π

Perhaps the best-known example of abstract mathematics is the calculation of the ratio of a circle's circumference to its diameter, as universally represented by the symbol π. Two related definitions are shown in Fig 4.3. On the left, we have a circle whose diameter is d; we can define the circumference as πd, a bit more than $3d$. On the right is shown the same circle, with its superscribed square; the area of the square is clearly d^2, and we can define π by saying that the area of the circle is $(\pi/4)d^2$, a bit more than $(3/4)d^2$. Abstract mathematics shows that the above two definitions of π are equivalent! The question now is: What is the value of π?

Figure 4.3: Related Definitions for π: (a) Using Lengths; (b) Using Areas.

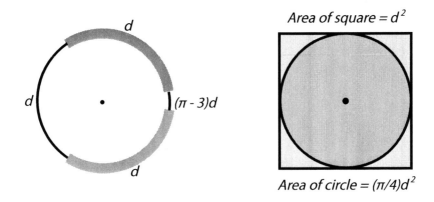

It can be shown *beyond all doubt*—not just beyond reasonable doubt, as in a courtroom, or to a very high degree of likelihood, as with the claims of science—that if one attempts to express π as a decimal fraction, the number of decimal places is not finite. Indeed, π has been shown by proofs in abstract mathematics to be *irrational*,[2] meaning that no finite fractional representation can be found in any base. Neither in base 10 (the decimal system), nor in base 17, nor in base 46,284,091, nor in any other base (positive finite integer). There are several proofs of this. Obviously, this is an elegant fact to anyone with mathematical sensibilities.

Even more interesting is the fact (again from abstract mathematics) that π is *transcendental*, meaning that it can never be a root of any polynomial equation of finite size that has natural numbers as coefficients. This all means that π is not easily calculated (to say the least) and, even worse, that there is no way to write out arithmetically what π is, except, well, to just write "π."

---

[2]"Irrational" is one of many unfortunate appellations from early mathematics that have persisted for so long that there is no hope for change. ("Pure" mathematics is another.)

Western readers may be familiar with the bible's implied value of π: A certain circular object was described in I Kings 7:23 as being ten cubits across and "a line of thirty cubits did compass it round about." This implies that $\pi = 3$, which is the correct value if one accepts accuracy to one significant digit.

When the author was in elementary school just over a half-century ago he was asked to use $\pi \cong 22/7$, which, as it turns out, is accurate to about four parts in 10,000. Accurate, in other words, to almost four significant digits—quite enough for a ten-year-old. Fortunately, the author was also told two other important facts: **(i)** that $\pi \cong 22/7$ is an *approximation*, and **(ii)** that the admitted error was quite tolerable "for present purposes."

Thus, from abstract mathematics we learn that **(a)** the two definitions in Fig 4.3 are identical; **(b)** π is irrational, so that there is no point finding an exact expression for π either in the decimal system, the binary system, or any other such system; **(c)** π is transcendental, so that there is no point in searching for π as a root of any polynomial equation, however large (but still finite) that has rational coefficients. We further learn from abstract mathematics that **(d)** there are many formulas for π based on infinite series, of which the best known is Leibniz's result:

$$\frac{\pi}{4} = 1 - \frac{1}{3} + \frac{1}{5} - \frac{1}{7} + \frac{1}{9} - \cdots$$

Unfortunately, this series converges so slowly that it takes 31 terms to get the error down to 1%, and 1,580 terms to get the answer to the same accuracy as the Grade 5 approximation ($\pi \cong 22/7$). There are better infinite series expansions to use; but surely the most important result is **(a)** above, since without it we would always suspect, but wouldn't know as an absolute certainty, that the two π's defined in Fig 4.3 are in fact exactly the same number. However, even here, the practical implications would not be large, since we would always know π_1 or π_2 to any desired accuracy by simple geometrical experiments.

Chapter 4: The Long Path

In 1950, the very early Eniac computer[3] was used to compute π to 2,000 decimal places. This is where our story takes on a truly giddy turn: using modern computer power, some say that the Chudnovsky brothers of New York have calculated the value of π to 2,260,321,363 decimal places. Consider the situation: it was already known that the number of decimal places for π was infinite; it was already known that a "circle" is a mathematical abstraction and that the number of (perfect) circles in the real world is zero; and it was already known that, even if a perfect circle *did* exist, no one could measure its diameter, area, or circumference with infinite accuracy. (Indeed, it must be very unusual when $\pi \cong 22/7$ is not of quite sufficient accuracy.) So what could motivate anyone to calculate π to 2,260,321,363 decimal places? Whatever it is, let's hope it's not contagious. Most especially, let's hope the Chudnovsky brothers (and similar folks who have much more time on their hands than they apparently need) were not being paid by, or using equipment paid for by, taxpayers! This example is extremely important because it simultaneously shows the uses and abuses of abstract mathematics.

To finish with a positive example, consider the following result, given by Archimedes (287–212 B.C.E.):

$$\frac{223}{71} < \pi < \frac{22}{7}$$

It has already been stated that the upper bound is in error by only four parts in 10,000; the lower bound is in error by *two* parts in 10,000. How practical! Archimedes is revered by abstract mathematicians even today as one of the greatest of all time, yet look at his mindset. He was comfortable with abstractions like π, yet sought results of utility like the one above. He also covered higher levels in the Long Path depicted in Fig 4.1. Archimedes was best

---

[3] *Eniac* (acronym for Electronic Numerical Integrator and Computer), the world's first electronic digital computer, had been developed by U.S. Army Ordnance to compute World War II ballistic firing tables.

known in his own time not primarily for his new mathematical ideas but because he developed scientific principles such as Archimedes Principle[4] and invented engineering devices such as the catapult (which was effective in the defense of Syracuse when it was attacked by the Romans). Archimedes trod much of the Long Path personally.

Example 2: Fermat's Last Theorem
As a second example of abstract mathematics, consider what is called the Last Theorem of P. Fermat (see box).

> **Fermat's Last Theorem (1630)**
> The equation $x^n + y^n = z^n$ has no solutions for integers $x, y, z > 0$, for integers $n > 2$.

Fermat scribbled this result in a margin where he also said he had found a "truly marvelous proof of this" (hence the "theorem" designation) but since many generations of the most talented number theorists (not to mention hordes of gifted amateurs) have since attempted to prove it without success it is now assumed that Fermat's proof was flawed.[5] However, the theorem *is true*, as proved by A. Wiles in 1994, just 364 years after its introduction. As a result in abstract mathematics, this proof is a great triumph of the human intellect. As to its practical importance, however, since Fermat's Last Theorem wasn't being used for anything anyway, the proof is pure "leakage," in the nomenclature of Figs. 4.1 and 4.2.

---

[4]Whether Archimedes ever actually shouted "Eureka!" [trans: "I have found it!"]—while running naked through the streets of Syracuse after having determined the proportion of gold and silver in a crown made for a king by weighing it in his bath water—will never be known, but it does serve to show his passion for the practical. Perhaps we should also give him credit for an early contribution to finance?
[5]Fermat was a lawyer by profession, although this is not meant to imply anything about his proof.

Chapter 4: The Long Path

Example 3: Guthrie's Four-Color Conjecture

For a much more recent example of abstract mathematics, consider the Four-Color Theorem (see box), which was first conjectured by F. Guthrie (who admitted that he had no proof) in 1852.

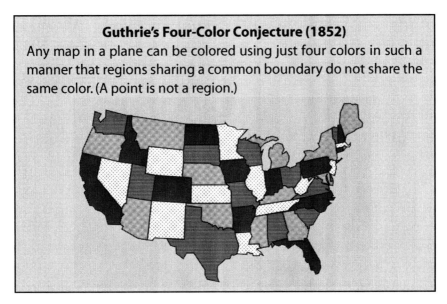

Guthrie's Four-Color Conjecture (1852)
Any map in a plane can be colored using just four colors in such a manner that regions sharing a common boundary do not share the same color. (A point is not a region.)

After several fallacious proofs were dreamt up, a correct proof was published by K. Appel and W. Haken in 1977, just 126 years after the problem was posed.[6]

The question of interest here is this: Does this theorem and proof have practical significance, or is this just more "leakage"? Put another way, does *National Geographic* want to redo all its maps as a result of the proof of this theorem? It seems unlikely, especially as the cartographers at NG were undoubtedly well aware of the four-color conjecture. It appears that the conjecture itself is actually more useful than the proof, although clearly the proof is a great achievement from the perspective of abstract mathematics.

When the best mathematicians cannot find any counterexample in 126 years (and these Everest-like problems in abstract mathematics

---

[6]When it takes 126 years to solve, it's a *problem*, not an *issue*.

do attract the best), practical people adopt an essentially scientific (not abstract mathematical) viewpoint. They say something like this:

> We are not *absolutely* certain that this conjecture is true, but of what, pray tell, in this life, *are* we absolutely certain? Meanwhile, we must color our maps, and we shall continue to use as few as four colors, if it suits us, until someone (perhaps one of us) comes across a case where four colors are not enough. We know that many gifted four-colorist abstract mathematicians are working this problem, and we'll raise a glass to the one(s) who eventually prove it beyond all doubt, but in the meantime we'll assume the four-color conjecture to be true. Indeed, this is probably the best-founded assumption any of us map-colorers will rely upon today, in any aspect of our lives.

Why should anyone wait for 126 years to color maps with four colors?

Example 4: Riemann's Hypothesis

In our final example, the proof is not yet in. G. Riemann contributed his celebrated hypothesis (or conjecture) in 1859. The hypothesis itself (see box) is more mathematical than our earlier examples, which may indicate a long-term trend.

Riemann's Hypothesis (1859)

Euler's "zeta function," namely,

$$\zeta(s) = 1 + \frac{1}{2^s} + \frac{1}{3^s} + \frac{1}{4^s} \cdots$$

defined for $R(s) > 1$, has all its nontrivial zeros on the line

$$R(s) = \frac{1}{2}$$

The trivial (unimportant) zeros occur at $s = -2, -4, -6, \ldots$

Chapter 4: The Long Path

One excellent source for recent discussion is Enrico Bombieri, whose paper[7] on this subject is somewhat sketchy about what this all means outside the monasteries of abstract mathematics. To prove the hypothesis, thus converting a conjecture to a theorem, is considered one of the "problems of the millennium," an appellation that at least suggests the anticipated schedule for its resolution.

With reminiscences of the history of π (see Example 1 above), Riemann's Hypothesis has now been vindicated for at least the first 1,500,000,000 zeros of the zeta function, which should give pause to those who might approach this hypothesis from a disproof direction.

Despite all the excitement,[8] we persist in asking: For the purposes of this book (i.e., for practical purposes), where's the beef? The application for Riemann's Hypothesis, when one can get any abstract mathematician to reflect upon it, is related to the age-old curiosity about the distribution of prime numbers. Further, if one knows the number of prime numbers less than a specified threshold, one has a critical leg up in the theory of *encryption*.

Encryption is undeniably critical to banking, internet commerce, national security, and much else. This connection has led some to opine that a proof of Riemann's Hypothesis would "bring the whole of e-commerce to its knees overnight." So there appear to be exceedingly hefty implications.

Conjecture On Utility Of Current Abstract Mathematics Work
Still, the author has misgivings and is not so convinced (though he has neither the time, the professional training—nor the inclination—to devote the many patient decades needed to prove his own conjecture). The conjecture is this: The *exact proof* of the Riemann

---

[7]E Bombieri: *Problems of the Millennium: the Riemann Hypothesis*, Institute for Advanced Study, Princeton NJ 08540. Proving Riemann's Hypothesis is considered by abstract mathematicians to be the most important of seven such mathematical problems for the new millenium.
[8]A $1M prize has been offered to the abstract mathematician (or anyone else) who proves Riemann's Hypothesis. This is an interesting shortcut to the commercialization stage of Fig 4.1.

Hypothesis—not just its demonstration for the first 1,500,000,000 cases—has the same degree of practical relevance as does knowing *exactly* the next billion decimal places in the value of π, or of knowing *exactly* the next billion non-solutions of the Diophantine equation referred to in Fermat's Last Theorem, or of finding out that the four-color conjecture holds for another billion planar maps. The associated usefulness, surely, has become vanishingly small.

Even if the conjecture in the last paragraph can be disproved in its "strong form"—that abstract mathematics now makes progress only on the irrelevant fringes—surely no one can make the argument that modern engineering students should spend anything other than a negligible quantum of time studying abstract mathematics. Perhaps a small amount of time would not be amiss for those students who choose the applied-science stream, in order to get some historical flavor, but certainly not for engineering students interested in design or management. To contrive curricula otherwise is just perverse and provides yet more evidence of the applied-science bias (Chapter 2).

The argument is sometimes made that all the wealth of effort expended on abstract mathematics over the many past decades is now happily available for all its applications that were unforeseen at the time. New applications, it is argued, though still unforeseen, will surely come along for the abstract mathematics now being slogged through. Even if much of this inventory goes eventually unused (the argument goes), the items that are exploited will more than validate the cost incurred in developing the entire body of work.

Unfortunately, evidence for this argument, used ubiquitously to justify salaries, research grants, and many other expenditures (primarily from taxpayers) are scant indeed. The oft-used (but threadbare) example of Boolean algebra, with its natural application to binary representations (thus anticipative of the development of modern computers), neglects to recognize that this mathematical machinery would have been speedily developed by the vast army of mathematically talented computer specialists, as and when needed.

The greatest mathematicians of old were much more than abstract mathematicians; they were driven by practical applications. We have already mentioned Archimedes in this connection. Newton invented a form of the calculus to study celestial mechanics. Gauss calculated orbits in his head. Euler; Poincaré; Lagrange; the list goes on and on. These individuals may be among the heroes of modern abstract mathematicians, but they were also from a much higher plane—they were motivated by realities other than the purely abstract in their brilliant discoveries.

4.2 APPLIED MATHEMATICS

Although it is academic heresy to say so (and probably politically incorrect as well), it seems that not much of *really practical* importance is being done in *abstract* mathematics. The dividing line between abstract and applied mathematics is rather blurry, and some applied math results do have some vanishingly small practical import (like a few more decimal places in π). Still, in the terminology of Fig 4.2, it's pretty much all leakage. The subject seems to be given a more-or-less free ride, based on its recognized ancient-history status as one of the bedrock foundations of modern civilization.[9]

Applied Mathematics

Not so, however, with *applied* mathematics (Fig 4.4). As its name suggests, applied mathematics is "mathematics with a purpose." It may not always be crystal clear just what that purpose actually is, but to earn this designation there must be some areas of practical importance, perhaps dimly perceived, that can arguably be aided by the developments in question. This is the first of the two distinctions between "abstract" and "applied" mathematics.

---

[9] It is noteworthy that, of the six Nobel Prizes, none is for mathematics.

Figure 4.4: Three More Stops on the Long Path—Applied Mathematics, Science, and Applied Science.

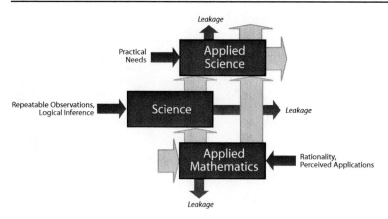

The second distinction follows from the first: Applications to real-world situations, including all of science, must involve some element of approximation. Abstract mathematics, by contrast, always eschews any approximation, however minuscule, as was demonstrated by the four examples in §4.1. This is certainly not to say that accuracy is a bad thing, but if the set of problems on which one can make exact mathematical statements includes virtually no situations of practical interest, and if one is asked to spend a significant portion of one's formative years studying this vanishingly small collection of curiosities, that is surely a worse thing.

Venn Diagram

Some additional insight into the relationships between applied mathematics, science, and applied science can be gleaned from Fig 4.5, which shows Fig 4.4 in a different way. This shows that there are important aspects of science that are not mathematics (see the discussion of science below); however, many scientific results use the language of applied mathematics to express laws or to deduce new results. Applied mathematics is often taught in the (abstract) mathematics departments of universities, but sometimes it is taught in a separate department, so to keep focused on applications and the approximate treatments that are unavoidable if realistic problems are to be addressed.

Figure 4.5: Venn Diagram for Applied Mathematics, Science, and Applied Science.

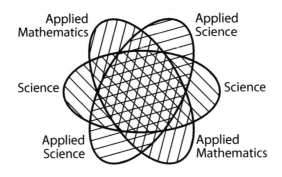

Applied science combines science and applied mathematics to utilize the results of science for practical use. Not all applied mathematics is bonded to scientific themes, however. One example of an activity that is applied mathematics without being applied science would be general statistical analysis; only mathematical principles (not scientific ones) are needed. However, if the results of such a statistical theory are applied to (say) the design and interpretation of scientific experiments, then this is clearly applied science. Another example of an area of applied mathematics that is not science is numerical methods—although these methods can also be used to help solve of applied science problems.

4.3 SCIENCE

Although the *results* of the scientific method are very important—and a great deal of engineering class time is spent on these results—the *scientific method* itself is even more important, yet is rarely examined properly in engineering school. In its fully developed form, science is very recent in human history and large parts of the globe do not really have it yet. Moreover, many areas of science still need considerable upgrading in their methodology.

Although science, as just defined, is a *process*, the word "science" is also commonly used to refer to the *results* of this process (for example, "medical science"). Perhaps engineering school curricula assume that the *process* has been adequately covered in high school, yet the general level of unscientific thinking (and even *anti*-scientific thinking) among the general population, and even among university graduates, is depressingly high.

> **The Scientific Method**
>
> The **Scientific Method** is based on *two* principal components:
>
> **Component 1:** Outstanding evidence (reliable, repeatable data from experiments).
>
> **Component 2:** Outstanding logic (rational inference).
>
> The **Scientific Method** also has several guiding principles, three of which are the following:
>
> **Guiding Principle 1:** If someone claims that something is true, it is the responsibility of the claimer to provide the evidence and logic to prove the claim. (It is not the responsibility of doubters to prove the claim false.)
>
> **Guiding Principle 2:** The more implausible the claim, the stronger the evidence required.
>
> **Guiding Principle 3: (Occam's Razor)** Of two explanations that equally fit the evidence, choose the one that is the simplest (i.e., that makes the fewest new assumptions).

Surveys reveal that, even among allegedly well-educated members of "advanced" western society, a nontrivial number believe in

Chapter 4: The Long Path

alien abductions, astrology, angels, devils, gremlins, water dousing, magic crystals, channeling from the dead, psychic phenomena, abominable snowmen, and the Loch Ness monster—the list would take many pages to complete. Yet none of these entities has the slightest reliable evidence and most have been disproved using proper scientific protocols.[10]

The process of acquiring scientific truth,[11] as shown in the box above, is long, arduous and difficult. It is always much easier to accept the latest fad, to believe merely what has been passed down from generation to generation, or—perhaps most enticing of all—to believe what feels good to believe. Most of our human brothers and sisters, given the choice between believing what is easy and attractive and believing what is difficult and disturbing, will choose the former. Perhaps we cannot really blame them, and perhaps in some areas of our own lives we yield to this same temptation.

Yet all the important advances in science have been made possible by very special individuals, relatively few in number, who, driven by their curiosity, their powers of analysis, their cerebral energy, their willingness to work hard for long periods of time, and—most of all—by their *intellectual honesty*, have disclosed relationships, insights and laws that describe how nature *works*.

From the picosecond (10^{-12} sec) flashes in modern spectroscopy, to the current studies of global warming on our human time-scale (10^9 sec), to the geologic crawl of Earth's tectonic plates (10^{-12} meters/sec), to the cosmological evolution of our universe over billions of years (10^{18} sec)—scientists have discovered amazing truths over **30** orders of temporal magnitude!

From atomic radii (10^{-12} meters), to the description of the killer Ebola virus (10^{-8} meters), to the size of the average engineer (2 meters), to the size of our own Blue Planet (10^7 meters), to the size

---

[10]One of TV's most notorious "faith healers" recently went to a leading U.S. hospital for treatment of his prostate cancer. There are lessons in this news.

[11]As stated repeatedly herein, when we say that a scientific result is "true," we do not refer to *absolute* truth, as is possible in mathematics, and as is claimed by many religious groups. We mean "true to a very high degree of likelihood."

of the observable universe (10^{26} meters)—scientists have provided enlightenment that is at times life-saving and at times life-fulfilling over **38** orders of spatial magnitude!

No other alleged source of information—not gut instinct, not folktales, not claims from the thousands of human religions, not intuition, and certainly not "what feels good" or "wishful thinking"—is even remotely close. Do all engineering students ever get a one-hour lecture on why science is the greatest achievement to date of the human intellect? If not, why not?

Certainly, science is, by its very nature, unfinished business. Moreover, it has come about only very recently in human history and it has a fragility that makes a return to the dark ages more threatening than we would like to contemplate. Notably, not all areas of science have matured in unison. For example, the basic laws of Physics seem, for all practical (engineering) purposes, quite mature;[12] the Medical Sciences are more recently gaining scientific stature (i.e., really saving lives and improving quality of life, not just pretending to do so); and the Social Sciences still have some distance to go before they conform to the rigors of the scientific method.

Scientific Method Example 1: Water That Remembers

We now consider briefly three examples of the scientific method. Together, they represent the power of the method in sorting the grains of wheat from the mountains of chaff, the nuggets of gold from the continual avalanche of dross.

The first example concerns one Jacques Benveniste, who in 1988 wrote an interesting paper in *Nature*, one of the world's most

---

[12]Though science has been disparaged by some as unreliable because new information "changes" old laws, the reality is much more positive and exciting. Old laws, if truly based on science, are still useful in the context of their original claims, but require extension and expansion to encompass a larger framework of applications as new data become available. (It is not easy to encompass the 30 orders of temporal magnitude and the 38 orders of spatial magnitude described above!) Thus, for example, the great Albert Einstein did not prove in 1905 that the great Isaac Newton was wrong in 1687 in his statement of the laws of dynamics; instead, he *extended* Newton's Laws to encompass much higher speeds and distances. To make the same point, Steven Weinberg, a Nobel laureate in physics, noted that we should still build a suspension bridge based on Newton's laws.

prestigious science journals. He claimed to have conducted elementary experiments with pure water as follows: first, he dissolved a small quantity of a substance in the water; next, he hyper-diluted the solution with additional pure water to the point where the mathematical probability was less than 0.5 that *even one molecule* of the dissolved substance was still present; and, finally, he made certain measurements that, he claimed, showed that the final water had different properties than pure water (even though that was what it was). Benveniste explained this result by deducing that the water "remembered" what had previously been dissolved in it.

Had this claim been promulgated through a lesser medium than *Nature*, it would have gone unnoticed, just one more item in the endless stream of unscientific offal that continually floods forth. But with the pedigree apparently given by *Nature*, many (real) scientists took note. They quickly recognized that what Benveniste was attempting was nothing less than a modern proof of the efficacy of *homeopathy*.

Homeopathy is a system of alleged medical treatments invented by S. Hahnemann (1755–1843) and is based on the notion that illnesses may be cured by administering vanishingly small amounts of a substance that causes the same symptoms presented by the patient. The theory is that symptoms are indications that the body is trying to heal itself. The best way to heal a patient is to make the symptoms worse, thus helping the body heal! So if a patient complains of stomach ache, the substance given would be something that causes nausea. Homeopathy obviously doesn't really make much sense and is counterintuitive, and no one has been successful in demonstrating these alleged effects scientifically; hence it is not part of modern medical science. Such a proof was exactly what Benveniste was implying (although he was careful not to explicitly mention homeopathy in his *Nature* article).

The scientific method went into action. First, Benveniste's data were re-examined by objective scientists. There had been counting errors, and statistical flimflams were also pointed out. Second, others tried to repeat the experiments in their own laboratories with disconfirming results. Today, no scientist takes Benveniste's work seriously and homeopathy is still back where it started—nowhere. Sadly, but not atypically for the species, Benveniste never recanted from his position, and in fact claimed more recently that he had invented "digital biology," by having recorded a signal stored in the water and then converting it into a computer file (which he e-mailed to sympathizers around the world). Even more amazingly, the signal can, according to Benveniste, be played back into a sample of pure water, which then takes on the properties of the original substance!

The author is unaware of any scientific time being spent on disproving these latest claims, and they were not published in *Nature*. Benveniste died in 2004.

Typically, the strange case of the "water that remembered" was prominently featured in national newspapers; its later documented exposure as just more flimflamery was not thought to be of such great interest.

Scientific Method Example 2: Energy From The Kitchen Sink
A quite different example (but with a similar denouement) also involves water, but in a distinctly different role. In 1989 two chemists at the University of Utah, S. Pons and M. Fleischmann, reported on a very simple home experiment: they just connected a pair of electrodes to a battery and immersed this contraption in a jar of heavy[13] water. They claimed to have witnessed fusion, the process that takes place on our Sun (whose center burns at 10^7 C). Pons and Fleischmann were working at a much cooler temperature, namely, the temperature of their kitchen sink. (Hot) fusion has been cre-

---

[13]Most everyone knows that (chemically pure) water is H_2O—two hydrogen atoms associated with one oxygen atom. Less well known is that the hydrogen atom (defined as having one proton and one electron) comes in more than one flavor—the technical term is *isotope*. Most (about 99.985%) H atoms comprise one proton and one electron, but a very few are "heavy" H atoms because the nucleus also contains a neutron as well. (Harold Urey discovered this in 1931, for which he won a Nobel Prize.) These unusual hydrogen atoms (or isotopes) are called *deuterium* atoms, with the symbol D used in place of H for clarity. The extra neutron makes D heavier than H and D_2O heavier than H_2O. Thus, D_2O is called *heavy water*, and it has atomic-scale properties that are somewhat different from H_2O, including being more obstructive to passing neutrons.

ated on Earth, not only in H-bombs, but at fusion research facilities—such as the JET laboratory near Oxford, England, where temperatures reach 10^8 C. Were this process completely controllable, it would offer the prospect of virtually limitless power. As of 2006, however, cold fusion methods do not yield more energy than is put into them, although (not surprisingly) unreplicable claims continue frequently to be made.

Scientific Method Example 3: The Cause Of Stomach Ulcers

Our third example of the scientific method has a happier ending for those who dared to dissent from the current orthodoxy at the time. Controversies in science are not solved by majority vote, or by issuing fatwahs, or by looking it up by consulting (and interpreting) some master book, or by appeal to some person who claims (or has been given) absolute authority. They are resolved by the scientific method.

> One's stomach must spend a lifetime in a highly acidic environment whose purpose is to dissolve large steaks in a few hours. It is noteworthy that the stomach does not digest itself! Surely we should at least suffer all our lives with intractable ulcers.
>
> Yet, in 1979, when R. Warren, an Australian pathologist, first observed the presence of strange, small, curved bacteria on a biopsy of the gastric mucosa of some of his patients, he was well aware that the cause of stomach ulcers in general—and duodenal ulcers in particular—was not a hot subject for medical debate. Everyone "knew" that ulcers were the result of irritation by certain foods and medicines, with genetic predisposition as a risk factor.
>
> In 1979 and earlier, ulcers could be fatal if they progressed to the perforation stage, wherein a hole was corroded right through the stomach or duodenal wall, spilling gastric contents into the peritoneal cavity. Thus, a perforated ulcer was a medical emergency, with surgery mandatory in a few hours if the patient was to be saved. Many died.[14]

---

[14] The author's first cousin, David C. Hughes, died on the operating table in 1961 from a perforated ulcer.

Shortly afterward, in 1983, pharma companies began to develop what were called Histamine H2-receptor antagonists, such as ranitidine (brands like Zantac). These medications were prescribed to combat the production of acid in the stomach at the source and saved many lives. It became one of the most prescribed medicines in the world, an indication of the importance of ulcers. About a decade later, another class of drug, called gastric-acid-pump inhibitors, such as omeprazole (brands like Prilosec), was introduced. With these pharmacological tools, ulcers became much less fatal.

Still, in 1979, these drugs were not yet available. Dr Warren had the scientific curiosity to follow up on his first inklings of a new hypothesis, and the next two years showed that the bacterium he had identified was indeed closely linked to gastritis. In 1981, Dr Warren joined forces with B. Marshall, an Australian gastroenterologist, and their ensuing partnership demonstrated the clinical significance of this bacterium, a new species now called Helicobacter pylori (or H-pylori for short). They demonstrated the association of H-pylori and peptic ulcers, particularly duodenal ulcers. Elimination of these bacteria resulted in healing of the gastritis; even better, the ulcers rarely recurred.

When Warren first presented his findings to the medical community, he was greeted with considerable skepticism—exactly as he should have been. The causes of ulcers were considered well understood. Besides, how could bacteria survive the acidic soup in one's stomach? (Answer: they secrete enzymes that locally neutralize the acid.)

Then the scientific method got into high gear. Other medical researchers attempted to replicate Warren's findings, and did. H-pylori infection is now known to be present in 90% of patients who have intestinal ulcers and 80% of patients with stomach ulcers. After treatment with a specific antibiotic, the bacterial infection usually disappears, as do the ulcers. The skepticism has melted away in the light of irrefutable evidence, and hardly a year now goes by that Warren is not awarded medals and honors of the most prestigious kind in the developed parts of the world.

Chapter 4: The Long Path

Millions of individuals have had their lives much improved—or even saved—by the application of the scientific method in just this one instance. Still, it is the *method itself*, not just a few of its myriad revelations, that deserves to be taught carefully to all educated people, not excepting engineering students. Science uses hard evidence and cool logic; this sounds so simple and sensible that it is astonishing how rarely it is rigorously applied.

Unsolved Problems In Science, And Their Application
Books ranging from *The End of Science* (by John Horgan, Broadway Books, 1997) to *The World's 20 Greatest Unsolved Problems* (edited by John Vacca, Prentice Hall PTR, 2005) opine on what's next for important scientific results.

Also of special interest to us, as engineers, is this question: Which of these new discoveries will have practical applications? We know that scientific findings can vary widely in their likelihood of application.

At one extreme we have areas like astrophysics, where the birth and death of stars cannot be applied to everyday life (although this knowledge arguably enriches our human understanding). At the opposite extreme we have fields like materials; it is scarcely possible to define where materials *science* ends and materials *engineering* begins. The cycle of explaining what is and then applying this new knowledge to what *might be* invented in applications is so tight that the two are almost inseparable.

4.4   APPLIED SCIENCE AND ENGINEERING
Applied science is part of engineering (see the pillars, Fig 1.6), though there are activities that qualify as being applied science that would not normally be called engineering. Thus we may think in terms of another Venn diagram (Fig 4.6), which can be compared with Fig 4.5.

Figure 4.6: Venn Diagram for Applied Mathematics, Applied Science, and Engineering.

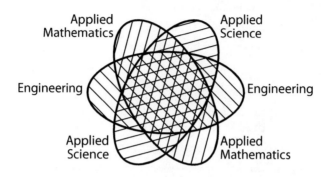

One aspect of interest is that there is a subspecies of applied scientist—what we might call the *engineering analyst*—that can be distinguished from the applied mathematician. This useful distinction is now explained via a series of examples.

Example 1: Structural Statics
Consider the general subject of the dynamics of structures. An abstract mathematician might begin with the following sort of equations:

$$\varepsilon(r) = \tilde{K}_{\varepsilon u} u(r) \; ; \; \sigma(r) = \tilde{K}_{\sigma \varepsilon} \varepsilon(r) \; ; \; f(r) = \tilde{K}_{f\sigma} \sigma(r)$$

where ε, σ, f, u and r are, respectively, the strain, stress, external force (per unit volume), displacement and position in a real (i.e., three-dimensional) structure; where $\tilde{K}_{\varepsilon u}$ and $\tilde{K}_{f\sigma}$ are first-degree, three-dimensional, linear partial differential operators; and where $\tilde{K}_{\sigma \varepsilon}$ is a linear algebraic operator. Evidently we can find the relationship between the cause (distribution of force, f) and the effect (distribution of displacement, u) as

$$\tilde{K} u(r) = f(r) \quad (1)$$

where we have combined the operators together thus:

$$\tilde{K} = \tilde{K}_{f\sigma} \tilde{K}_{\sigma \varepsilon} \tilde{K}_{\varepsilon u}$$

Chapter 4: The Long Path

It may not jump out at the reader using this kind of notation, but Equation (1) represents a very large system of linear partial differential equations (PDEs). Written out in full detail for, say, a 747 aircraft, it would take many, many pages to display (each strut might take a page or two), but of course no one would do that because in that form it can't be solved.

Strangely, an abstract mathematician doesn't particularly care that the complete solution can't be found (meaning: extracting numbers of engineering significance); he or she is more interested in proving that the solution to (1) exists, that $\tilde{K}$ is a symmetric positive-definite[15] operator, and things of that sort. These are helpful to know, but the aim for an engineering analyst is to get numbers, not for some case that is so simplified as to be meaningless, but for the real structures that are presented to her.

In contrast to the abstract mathematician, the applied mathematician wants numbers. For many, many years, he didn't know how to get them. One of two things happened: either **(a)** the only concrete example[16] actually given of a (so called) structure was a long slender boom; or **(b)** some sort of finite difference scheme was applied, complete with "coefficient molecules," extrapolations, interpolations, and all the rest. But mostly, they solved the slender boom again and again. Students please note: It was *engineering analysts* who developed a practical approach to this class of realistic problems, as we shall presently see.

---

[15] Assuming no rigid displacements are permissible. If they are, we must settle for positive *semi*-definiteness.

[16] This long slender boom (sometimes called the Euler beam) has one dimension, not three, and it has just one parameter type, a stiffness parameter, and even this one is a scalar and is constant along the entire length! It is the world's simplest structure and no structural engineer can earn a living being interested in it. The applied mathematicians who use this example and only this example of whatever method they are discussing pretend that it's just a little example at the end of their paper and that time does not permit an illustration of the more fulsome merits of their work. However, if one pins them down later at the bar, they are forced to admit that they don't know how to apply their new dynamite method to any structure that is more complicated! These people are trapped half way between abstract and applied mathematics; many of them are would-be (abstract) mathematicians, wishing they could avoid the bother of solving any practical mathematical problem that is more complicated than the completely trivial long slender boom.

Example 2: Structural Dynamics

Our second example is a continuation of the first: we now glance briefly at structural dynamics, which means that we must add time as an independent variable, and we must also be given the structure's mass properties. From d'Alembert's principle, we can extend the force field to include inertial forces:

$$f(r) \Rightarrow f(r,t) - \rho(r)\ddot{u}(r,t)$$

where ρ is the mass density at a point in the structure represented by r. This can be written in the alternate form

$$\tilde{M}\ddot{u}(r,t) + \tilde{K}u(r,t) = f(r,t) \quad (2)$$

where $\tilde{M} = \mathbf{1}\rho(r)$, $\mathbf{1}$ is the unit dyadic, and overdots denote differentiation with respect to time. Note that equation (2) is now a potentially huge (if written out in detail) system of partial differential equations in *four* independent variables, with spatially-variable coefficients.

Applied mathematicians did not know how to solve (i.e., get numbers from) this equation, for structures of realistic complexity, until engineering analysts showed them how.

Approaches To Approximation

Even though applied mathematicians know that getting numbers should be the end objective, they are still often obsessed with the goal (though an unnecessary and impossible one) of complete accuracy. Engineering analysts, by comparison, feel more comfortable with making assumptions, particularly when the associated errors are understood and can often be reduced by paying for more computer power.

The aversion to approximation in much of applied mathematics has one of two consequences in those relatively rare situations when practical (physical) problems are contemplated. Either no progress (or essentially no progress) is made, or the approximations are bunched up at the modeling end, so that the solution can be claimed to be "exact."

Chapter 4: The Long Path

This strategy is illustrated in Fig 4.7. In the applied mathematics approach, the geometry is first simplified to that of a circle, an ellipse, a square, a rectangle, etc. Then key physical parameters are assumed to be zero or, at most, constant. Is the structure dish-shaped? Then flatten it. Does the structure have non-uniformities? They'll have to be ignored. Do we know the significance of "end effects"? Heck no. Does the environment change the properties? Probably, but who knows how to do that? Somehow the 747 aircraft becomes a very simple structure indeed.

Figure 4.7: The Shell Game of Approximations.

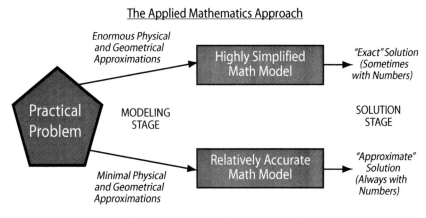

The resulting (highly simplified) math model sometimes has a solution—usually expressed in terms of what are called higher functions, that are, by definition, the solutions of the (highly simplified) mathematical problem. This solution is referred to by its perpetrators as the exact solution, although clearly the claimed exactness was preordained by all the simplifying physical and geometrical assumptions made at the earlier (modeling) stage, not to mention all the subsequent numerical approximations in the evaluation of the higher functions.

101

This all might be forgivable and written off as mere semantics were it not for the air of moral superiority radiating from the mathematicians who are offering their so-called exact solutions. In fact, of course, both models are approximate and both solutions are approximate, with the important distinction that the "finite element" approach, historically developed by engineering analysts, not applied mathematicians, usually produces a result that is much more accurate than is the solution based on applied mathematics approaches. The engineering analyst's approach—in our current example the finite element method (FEM) solution—though derogated as "only approximate" by many applied mathematicians, can solve structures of arbitrary complexity to any desired accuracy (see below), something no one has ever done using the methods recommended before the FEM was developed.

Applied mathematicians are also prone to saying things that are just plain silly about their "exact" solutions. "Your solution does not have an infinite number of vibration modes," they will say, "but mine does." This raises an interesting question: How many modes does a real structure (not someone's over-simplified equation) actually have? The correct answer is zero. The curiosity of modes is made possible only by the assumptions of linearity (both geometrical and constitutive) and by disregarding the second law of thermodynamics, which guarantees that there will, in reality, though perhaps not in somebody's bereft equations, be dissipative processes. Modes represent an extremely useful concept but they are characteristic more of *equations* than of *structures*.

Furthermore, when PDE enthusiasts fulminate that their models have an infinite number of modes, they should be made to realize that this is nonsense. More modes than the number of molecules in the structure? Or more, even, than in the known universe? The prediction of an infinitude of vibration modes is not a strength, but just another lamentable weakness, of the simplified PDE-continuum approach.

Chapter 4: The Long Path

Engineering Analysts At Their Best—the Finite Element Method
The finite element method (FEM, for short) was developed by engineering analysts, like the late John Argyris and others, who were being asked to do stress analysis and modal evaluation for structures of increasing complexity—and there was essentially nothing to help them in the task. (As an example, consider the structure shown in Fig 4.8, recently studied at McMaster University, in Hamilton, Ontario.) There was Euler's long slender beam, and some similar oddities, and a book had been written on the uniform, thin rectangular plate. Nothing much of practical help there. In fact, not only could the PDE formulation not be solved (except in pathologically simple cases), it could rarely *even be written out* at a level of detail befitting the real structure! And the finite difference molecules often recommended were unwieldy and of uncertain accuracy. Something had to be done, and it wasn't getting done by mathematicians.

Figure 4.8: Finite Element Model (right) of the Canadian Parliamentary Library (left).

In the 1950s and 60s, structural analysts decided to start from scratch, stressing (pun intended) the physical relationships rather than the mathematical niceties. Calculus and limits were discarded and the interactions between a very large number of very small structural elements were written out. Since these relationships were linear, the *stress distribution* problem was transformed into a problem in linear algebra, with which the development of computers was ideally suited to assist. Further, using vibration modes (a finite number but more than enough to do the job), structural *dynamics* problems were also transformed into problems in linear algebra.

The character of the idea can be sketched by continuing with the analysis typified by Equations (1) and (2) in the examples above (although this notation would not have been used by the pioneers of the FEM). Using a Ritzian expansion like

$$u(r,t) \cong \Psi(r)q(t)$$

where $\Psi(r)$ is a rectangular matrix of shape functions and $q(t)$ is a column matrix of generalized independent displacement variables, and applying variational principles of dynamics, one arrives at a structural model of the form

$$M\ddot{q}(t) + Kq(t) = f(t), \quad M = \int_E \Psi(r)^T \Psi(r) \rho(r) dV,$$

$$K = \int_E \Psi(r)^T \tilde{K} \Psi(r) dV$$

where E is the domain of the elastic structure.

Finite element methods are now used in fluid flow, thermal analysis, electromagnetic analysis, and virtually every corner of applied physics. They include many details of geometrical and physical complexity and can be made increasingly accurate by increasing the number of nodal variables (represented by $q(t)$ in the above structural example).

Although the above examples are understandably from the author's particular research experience, similar examples abound in other areas of engineering interest.

4.5 NEXT STEPS IN THE LONG PATH

Turning again to the Long Path depicted in Fig 4.1, we see that the areas of abstract mathematics, applied mathematics, science, and

Chapter 4: The Long Path

applied science (including engineering analysis) have been discussed on the preceding pages in this chapter. The remaining areas are shown in Fig 4.9.

Figure 4.9: Steps Remaining in the Long Path.

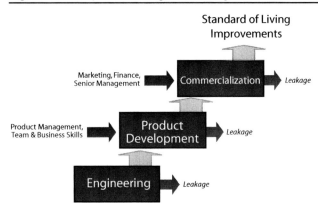

No further comments will be made in this chapter on the areas of engineering that are not applied science. This has already been covered in Chapters 1 to 3, particularly Chapter 1. Product Development will be briefly examined in Chapter 9; Marketing in Chapter 5; Finance in Chapters 8 and 12; and Senior Management is highlighted in Part III (Chapters 11 to 13).

Epilogue: Several months after this chapter was written, Australians Barry Marshall and Robin Warren were awarded the 2005 Nobel Prize in medicine "for showing that bacterial infection, not stress, is to blame for painful ulcers in the stomach and intestine." The scientific method (§4.3) has now come to its final conclusion in this matter.

CHAPTER 5

Marketing Concepts

Giving Customers Pleasure

For many years the author gave a course on technology transfer to fourth-year engineering students. The idea was to help these young people better realize all the ways their work over the first decade or so of their professional life could be useful—possibly even crucial—to the organizations that employed them. Most senior undergraduate students (or at least the ones in this course) were hungry to find out why their work would be important, and how it would benefit not only their own careers, but society as a whole. It is the author's opinion, developed from these and other similar experiences, that engineering students are more tuned into subjects like this than they were even a decade or two ago (and probably more so than many of their professors).

Since the technology transfer course was an optional course (and most students were not permitted by their home departments to take it, even as an option), it was scheduled in the evening. This gave the author the opportunity to speak at length with many of these students over the years, after class, about their interests, fears and enthusiasms as they prepared to enter the workforce. This led, in turn, to an informal data collection exercise, in which the author was able to learn what these students knew and didn't know about the world of business. The conclusions, in short, were **(a)** engineering students are frighteningly innocent of most aspects of the commercial world (unless they have some special information source, such as a near relative), and **(b)** the greatest area of vacuity, relative to its importance, is undoubtedly marketing.

True, the word *marketing* has some unsavory nuances for all of us. The ubiquity and intrusiveness of modern commercial messages can leave one with a bad taste in one's mouth. "Would you buy a car from this man?" is the caption on so many cartoons and photos—always an unfortunate put-down of salespeople[1] everywhere. Especially, and unfortunately, it is often true that "engineering types" and "marketing/sales types" typically bear each other the same natural affection as do cats and dogs. Engineers deal in hard-as-rock reality; marketeers create images. Salespeople make promises; engineers have to deliver. This friction is indeed inopportune because, unlike dogs and cats, the technical and sales species really do need each other, and rather desperately at that, in any modern business.

Based on these conversations with engineering students and his own understanding of business, the author has made it a point to stress the importance of marketing and sales to every young engineer with whom he comes in contact. The response is usually disbelief, followed by a period of introspection and cognitive rearrangement, completed by a grudging acceptance (at minimum) that marketing and sales may well be the single most important part of business.

5.1 BASIC MARKETING CONCEPTS

The state of innocence with respect to marketing (which, in our usage here, includes sales) explains the presence of this chapter in this book. Everyone talks about the importance of the bottom line, but

---

[1] Very few organizations call their salespeople "salespeople." There are dozens of euphemisms, up to and sometimes including "Vice President."

Chapter 5: Marketing Concepts

there would be no bottom line were there no top line. A business cannot survive without revenue. Some startup biotech firms and pharmaceutical companies do last a long time without sales while the product is being developed; they exist on early investment for an extended period, much as native pearl divers can dive underwater and hold their breath for several minutes—the other 99.99% of us can't last one minute—but eventually they must surface to breathe or they die. Similarly, companies must have sales to survive, and robust sales to grow and thrive.

Figure 5.1: No Customers → No Business.

It is in this sense that the customer[2] is so important to the business and should never be forgotten, even in the R&D laboratory. This is the most important lesson in this chapter. The rest is merely supportive material.

This section will review some of the most basic ideas in marketing. If the student has had a marketing course—which seems unlikely—much of this material will be old territory. It is hoped, however, that there will also be some fresh insights even for such readers.

---

[2]Low-brow outfits have *consumers*; medium-brow outfits have *customers*; and high-brow outfits have *clients*. We'll use the neutral word "customer."

The three major concerns in marketing (other than the actual sale itself, which will happen when satisfactory answers are found to the three concerns) are shown in the accompanying box:

> **The Three BIG QUESTIONS in Marketing**
> 1. What is my market?
> 2. What is my product?
> 3. How do I reach my market with my product?

Note the order of the first two questions. The natural order for someone training in engineering is more like this: "What is my product?" And then (if any human or financial resources are left over) "How can I sell this stuff?" However, this is the reverse of the better order shown in the accompanying box.

Technology Push vs. Market Pull

This ordering of questions is closely related to the technology-push market-pull issue (Fig 5.2). Engineers, applied scientists and technologists go naturally for technology push, even though this choice is more through inadvertence than a carefully chosen strategy. They find themselves with some technology (that's what they do) and then they decide to sell it. If some resources are available, they improve the technology, and this passes for marketing. They tend not to spend much quality time thinking about who might want to buy it, nor about finding them and asking what features might interest them and how much cash they would be prepared to spend in order to procure the product.

Figure 5.2: Which Is the Best Strategy?

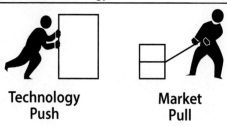

Chapter 5: Marketing Concepts

Market pull, by contrast, starts with the market. Of course one must have *some* idea of the product one wishes to bring to the market; otherwise, there is no way to identify the market. But the point is that the market (the customer) is brought into the process much earlier and much more often if market pull is used.[3] Even before the serious design stage, real markets are identified and real customers in those markets are contacted. Their counsel is sought on which features are most desirable, what prices would produce how many sales (the demand curve) and other similar (and critical) information. This process is generally known as market research,[4] and should be a strong influence on design, functionality and styling, not just an afterthought.

Segmentation

Once one has identified a (theoretical, or aggregate) market, meaning a set of potential customers, one looks at these customers more carefully. They are not all identical and their differences can be exploited to target them (satisfy their needs) more precisely. This process, called market *segmentation*, is illustrated in Fig 5.3.

Figure 5.3: Segmentation of Market.

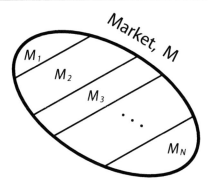

---

[3] One aphorism states the case thus: "Markets don't write checks; customers do!"
[4] The author attended a commercialization conference recently where speaker after speaker, all leaders in their high-tech companies, used the well-known phrase "R&D." It eventually dawned on this listener that it was *market research* and *product development* that was being referred to. They were using market pull, while the "other R&D" often leads to technology push.

There are many segmentation strategies available to the marketing innovator. Any discriminating characteristic (or better, a unique combination of characteristics) can potentially be used; the marketer's craft seeks to choose the best of these. If one's customers are individuals (not organizations), the following are examples of segmentation strategies and parameters: *demographic* (age, sex, culture, economic or social class); *psychographic* (lifestyle, attitudes, values, or other personality traits); *geographic* (region, climate, growth rate, population density); and *behavioral* (usage patterns, price sensitivity, brand loyalty, and benefits sought). Market research will help to suggest the most fruitful approach for a particular product type.

Positioning

One segments one's market, and one *positions* one's products created to respond to those segments. For example, one might create a product for each segment. Fig 5.4 illustrates this process for automobiles. The two positioning parameters used happen to be degree of sportiness and degree of luxury.[5] This sort of figure is sometimes called a *positioning map*.

Figure 5.4: Positioning of Four Automobiles.

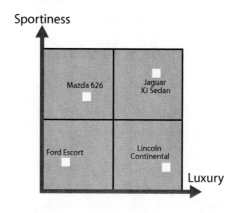

---

[5]The author acknowledges the assistance of some knowledgeable senior engineering students in creating the examples in Fig 5.4.

Chapter 5: Marketing Concepts

Another way to look at positioning is in terms of the supply and demand curves for each product, which intersect at their equilibrium point. Suppose our company has four products in a certain market, one positioned in each of four segments. The price point for each product (one for each market segment) is ideally at the intersection of the supply and demand curves (Fig 5.5). The four intersection points suggest a sort of equilibrium curve for the company's products in this aggregate market.

Figure 5.5: Supply/Demand Characteristics of Four Products, One in Each of Four Market Segments.

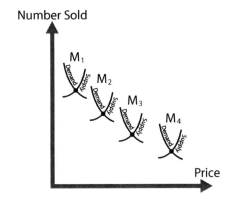

Differentiation

One segments one's market, and one positions one's products. One also strives to *differentiate* one's products from one's competitor's products in the same market segment. Sometimes this differentiation is just smoke and mirrors, as when our over-the-counter drug has an active ingredient that is identical to our competitor's offering, but pharma companies try to juice up their pills with more attractive packaging. Other times the basis for differentiation is more substantive. We can also plot our competitors' products on our position maps to gain further insight. As a final remark, the process of differentiation is closely related to the notion of *branding*, in which special names, symbols, designs, etc., are used to differentiate products and even whole companies. The decades-long battle between Coke and Pepsi is a fine example of branding and differentiation.

Niche Marketing

Small companies—especially startups—are always on the lookout for small clusters of low-hanging fruit that have not been picked by the larger firms (Fig 5.6). If a $1B firm wants to grow by 10% over the next year, it has to find an additional $100M in revenue. New business propositions that generate, say, $4M in revenue, even if quite profitable, are not worth the trouble to the big players. The calculation is quite different for an $8M firm.

Figure 5.6: A Niche Market—Small but Profitable.

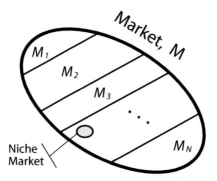

Just as there are diminutive species in the natural ecosystem who find tender morsels overlooked by, or of no interest to, the great predators who roam doing their big deals, in the business ecosystem there are similarly smaller players who can make a good living on niche markets. All the principles of marketing still hold true, although they have to be fine-tuned to the reality of smallness.[6] If the small company is very successful, and finds itself growing large to the point that its niche markets no longer satisfy its voracity, then that is a problem all small companies should have.

---

[6]RE Linneman, JL Stanton, Making Niche Marketing Work, McGraw-Hill, 1992.

5.2 MORE ABOUT PRODUCTS

Having now established the basic vocabulary and concept tools needed to discuss marketing, we move on to examine some further issues, with special emphasis given to the marketing and sales of engineering and technology products.

What Is A Product?

As with many words that seem familiar and uncomplicated, "product" has some hidden nuances that are worthy of comment. The basic meaning we shall use is recorded in the accompanying box. This formula is not algebraic; it simply means that virtually all conceivable products have both a *good* component, and a *service* component.

> **Definition of *Product***
> Product = Good + Service

Of the pair, "service" has the same meaning as in everyday conversation, but "good" is somewhat more puzzling and is meant to convey the sense used by economists, lawyers and accountants. Here are some definitions of *good*:

> **From a top textbook[7] on marketing:** *"A set of tangible physical attributes assembled to an identifiable form to provide want-satisfaction to customers."* Presumably "providing want-satisfaction to customers" is in practice extraordinarily similar to "satisfying customers." The key words are "tangible" and "physical," and the latter obviously includes the former.

> **From a top textbook[8] on business law:** *"A tangible thing that is movable at the time of its identification to the contract."* Note the agreement on tangibility, plus the natural reference to a legal contract.

> **From another top textbook[9] on business law:** *"... goods derive their value intrinsically, that is, simply because people want the goods themselves, for the utility or satisfaction they furnish."* This

---
[7] MS Sommers et al., *Fundamentals of Marketing*, McGraw-Hill Ryerson, 1992.
[8] HR Cheeseman, *Business Law* (Fifth Edition), Pearson Prentice Hall, 2004.
[9] JE Smyth et al., *The Law and Business in Canada*, (Sixth Edition), Prentice Hall, 1991.

definition, if a tad circular, is refreshingly to the point: people want goods because they are intrinsically satisfying.

From an on-line dictionary: In law, "a comprehensive name for almost all personal property as distinguished from land or real property."

The origin of the word "good" (with the meaning we are discussing) is clearly ancient, which explains the reluctance of the deeply historical professions to replace it, even though it has fallen out of general use. Even two generations ago people might think nothing of tripping over to the "dry goods store" on a sunny afternoon, but the word has all but disappeared from modern non-professional usage.[10]

Accountants are also sensitive to the word "good," because GAAP[11] refers to it often. The inventory of "finished goods" is but one example. If lawyers and accountants use this word with precision, it is worth learning, quite apart from the slightly more general business nomenclature.

> **Definition of a *"Good"***
> [for engineering students only]
>
> If a candidate good (CG) can be thrown out a tenth-story window; and if, having been thus thrown out, the CG will fall under gravity to the ground below; and if this falling CG might hurt someone if it lands on them; and, finally, if someone unhurt below but in the vicinity would likely take a good look at this CG to see what residual value it had—then the CG is a good.

With the strange word "good" now understood and behind us, let's return to the word "product." Most readers might at first think that "product" is just the new word for "good," with the added emphasis that

---

[10]One exception perhaps: A mugger with high verbal IQ might say, "OK, Gimme da goods!"
[11]GAAP = Generally Accepted Accounting Principles. See Chapter 8.

it was produced by human enterprise of some sort. Thus, while "good" tends to emphasize ownership, "product" tends to emphasize process. However, as the definition of "product" above indicates, there are two points of importance: **(a)** "product" for us will mean either "good" or "service," or both, and **(b)** the pure case of "good" or "service" is virtually nonexistent. It is difficult to think of products that don't have both some tangible value and also some service component.[12]

> For example, a middle-aged gentleman purchases a high-end automobile. Certainly the "good" is plain to see, as are all the toys and options on board. Still, if the dealer does not honor the maintenance agreement, the 24-7 roadside assistance and numerous other service items, this "product" will be considered a failure.
>
> Or, as a second example, a young woman who has just been promoted to director of strategy at a mid-sized firm decides to ask for the written advice of two consultants on two particularly vexed issues. The consultants deliver their reports, complete with paper copy, CD version and a box of reference material. There are both service and "good" components to this transaction.
>
> From the point of view of law, the car may be classified as a good, and the consulting report a service, but from a business standpoint both have both good and service components.

Some writers refer to the *complete product* or the *whole product* so as to include the service component. We shall just say product. This includes the good, service, color and styling, warranty, brand, reputation, price discounts and a long list of other attributes. Customers don't just buy something; they have an experience—and it had better be a good one.

Price

Marketing instructors often speak of "the four p's," these being product, positioning, price, and promotion. We shall not discuss promo-

---
[12] This phenomenon of a combined value proposition is called "bundling."

tion (which includes advertising, among other things) but a word or two about price is of interest. Everyone has a personal interaction with many prices each day, and, further, price has a quantitative ring that makes young engineers feel they understand what is going on.

The simplest pricing question for the marketing department is "What price should we charge for this product?" The answer to the question is not straightforward because simply raising prices might lose customers (the demand-curve effect). It will also lose market share and the economies of scale. *Lowering* prices, on the other hand, calls to mind the joke about the marketing neophyte who said he was losing money on all the individual items but was making it up on volume.

The way that virtually all businesses address this issue is to not answer the stark one-price question directly, but to ask a different one instead: "What should our *pricing policy* be?" The answer, as it turns out, is not a single number, but a whole matrix of numbers, each with many conditions attached. If one goes into a fast-food restaurant, chances are one will be asked if one has the latest discount coupons. This lowers the price for couponed customers. In fact the whole coupon industry—pages and pages of them from many sources—are a way of adjusting the price (down) for the segment that is willing to mess with coupons.

> Or, try this mental experiment: Stand up at the front of a commercial airliner (after safe takeoff) and ask everyone on board what they paid for their seats. We know that first-class travelers pay more than business class, who in turn pay more than economy (or tourist) class. This much is just segmentation. But how do we explain how all seven passengers in the same row each paid a different fare? There may not be two identical fares in the whole cabin.

Through these and many other examples, we begin to realize that there is no one price for anything. The technical term in economics for this process is *capturing the consumer surplus* (Fig 5.7) and the devices for arranging this capture are myriad. If only one price were chosen, p^*, then the demand curve in Fig 5.7 informs us that the number sold at that price would be n^*. The revenue is simply n^*p^*, with the area interpretation shown. The problem with this one-price policy is that there were several customers willing to pay $p^* + \delta$, and several more willing to pay $p^* + 2\delta$, etc. If we only charge p^*, we lose out on all this extra revenue.

Figure 5.7: Capturing Consumer Surplus.

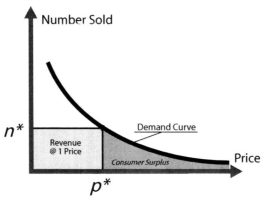

Therefore, the object of the pricing strategy is to capture all those missing revenue dollars; this revenue "surplus" is also shown in Fig 5.7 as the area under the demand curve for $p > p^*$.

Value

Value is a hard thing to pin down. It is essentially experiential and based on perception. Small wonder that descriptions tend to wax eloquent and vague almost to the point of frustration. One reasonable definition, what might be called the "barter" definition, is to say that two items have the same value if they could be exchanged at no loss. This definition even defines an experiment for measuring the value of a product. It still leaves a lot to individual assessment, however; and it is also somewhat circular, giving the value of one thing in terms of another (unless that other thing is a wad of greenbacks).

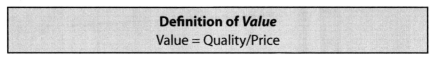

Definition of *Value*
Value = Quality/Price

Some benefit can be gained, perhaps, from the definition in the accompanying box. The higher the quality for a given price, or the lower the price for a given quality, the higher the value. If there were numerical values for quality and price, the arithmetic would be simple enough, but we have just seen above that price is hard to pin down, and quality is, well, qualitative. However, if a system were agreed to for quantifying quality, value would be numerically known and comparable.

Quality

In the last decade, the quantization of quality has grown by leaps and bounds, and having a nice definition for value is one of the benefits. The most successful approach has been dubbed six-sigma, which we shall write simply as 6σ. Based on the Gaussian (or *normal*) distribution (Fig 5.8), we recognize σ as the standard symbol for standard deviation—in this case the deviation from perfection, part of the precisely defined specifications for the manufacturing parameters.

Suppose the specification for error is δ, meaning that the customer is completely satisfied with errors within $\pm\delta$ of the target. If the statistics of our manufacturing or other product-preparation operations are such that $\sigma=\delta$, this means that 15.87% of operations produce a result unacceptably high (and a further 15.87% unacceptably low); altogether, 31.74% of operations produce a result unacceptable to the customer—an incredibly lax standard. No one can stay in business with quality this poor, except perhaps a hotdog vendor at a rock concert.

Figure 5.8: Gaussian Distribution of Operating Results, for $\delta=\sigma$, 2σ, 3σ.

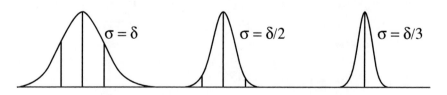

Chapter 5: Marketing Concepts

One can raise the bar somewhat[13] by making ±2σ the standard for quality instead of ±σ. Thus, σ= δ/2 Now 95.4% of operations are acceptable to the customer.

With σ= δ/3 99.7% are acceptable, and so on. (This is already too small to see in Fig 5.8.) Continuing in this vein, so-called 6σ means that σ= δ/6 there are only three or four defects per million operations.

At this point the reader might be saying "Whoa! This exquisite quality would be nice to have, but surely it's prohibitively expensive." Actually, no. In fact, if one's competitors are achieving this level of quality, one's own business cannot afford *not to*. Even if one's competitors have not reached this lofty level (yet), there is still ample incentive to improve quality because—and this is the key point—if done correctly, high quality is actually *cheaper* to produce, *overall*, than low quality. (We are talking here of modern manufacturing with all the technology that can be brought to bear.) Figure 5.9 shows some of the reasons. In the long run, the company actually *saves money by doing things right the first time*. As with icebergs, most of the cost is hidden.

Figure 5.9: The Obvious Costs (and Many Hidden Costs) of Poor Quality.

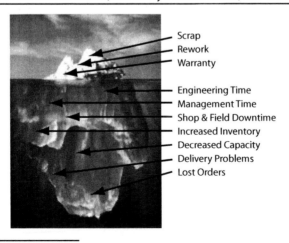

Scrap
Rework
Warranty

Engineering Time
Management Time
Shop & Field Downtime
Increased Inventory
Decreased Capacity
Delivery Problems
Lost Orders

---

[13]Or, if the metaphor is the limbo, we lower the bar somewhat.

This miracle of quality can be achieved only with a commitment on the part of every person in the company to this level of quality. The literature contains detailed and inspiring tales of how this was done.[14] In the late 1980s and early 1990s, Motorola went from $\delta= 4\sigma$ to $\delta=5.5\sigma$, saving US$2.2B on the bottom line, after deducting all the costs associated with reeducating their entire workforce. For a large company, this learning process takes five to ten years, even with great leadership and turned-on personnel.

The Motorola experience caught Jack Welch's attention at General Electric (as potential $2.2B savings were wont to do with this dynamic CEO) and he got GE all the way to $\delta=6\sigma$ by 2000, again with startling results (savings were US$750M and US$2B respectively in 1998 and 1999). The UGBO for these developments is stated in the accompanying box.

> **UGBO (Unerring Grasp of the Blindingly Obvious)**
> If you see a path to make your customers happier, your employees more enthusiastic, your products more economically manufactured, and your margins higher—and if it is legal and ethical—**take it!**

In effect, the organization-wide commitment to quality becomes not only a massive shift in the cultural paradigm, it is a major capital project with serious investments over several years, but with a great return on that investment for many more years to come.

Green Belts, Black Belts, And The Hidden Factory

The achievement of $\delta=6\sigma$ quality performance requires nothing less than a minor brain transplant on the part of everyone involved,[15] though naturally of some more than others. There are two major strains of resistance to the conversion from $\delta=?\sigma$ to $\delta=6\sigma$—and notably neither strain infects engineering students. First, there is the

---

[14]For example, D Smith et al., *Strategic Six Sigma: Best Practices from the Executive Suite*, Wiley, 2002.
[15]T Pyzdek, *The Six Sigma Handbook: A Complete Guide for Greenbelts, Blackbelts, & Managers at All Levels*, McGraw-Hill, 2001.

Chapter 5: Marketing Concepts

resistance to change, explained in the Chapter 7 (§7.2); students are too young for that. Second, the statistical basis for 6σ makes it a trifle forbidding for many, but engineers have had at least one course on probability and statistics, so the 6σ process should stimulate them, not frighten them.

> As an aside, and to connect with Chapters 2 and 3, some readers may wonder whether the author is being consistent. Chapter 2 spoke of an applied science bias in engineering schools and Chapter 3 challenged the prevailing view that *all* engineering students, even the majority who will become engineers (not applied scientists), should be asked to study so much mathematics—both in math courses *per se* and also in many other courses that are really applied math courses in thin disguise. How, then, can a probability and statistics course now be good?
>
> First, all engineering students should take some mathematics; that is obvious. The questions are (or should be): How much math, and what type of math? For students who want to be engineers, there should be far less emphasis on calculus, that darling of the applied scientist, but rarely (if ever) used by practicing engineers. This means fewer calculus courses, and less calculus in all the other courses, for aspiring engineers. Calculus doesn't just collect on teeth; it also seems to collect on engineering curricula.
>
> Furthermore, were the author asked to choose between one course in calculus and one course in probability and statistics, the decision would definitely be in favor of the latter. Not only is it far more useful in engineering, it also creates a mindset and understanding that help the student deal as an educated citizen with a multitude of issues.

In addition, young people are idealistic, and 6σ is definitely idealistic (though realizable). The following is instructive of a great deal more than quality improvement:

> Four processes have been identified as crucial to the achievement of 6σ: measure, analyze, improve, control. To imbue all staff with this understanding requires that an internal school be set up within the

company. (Some of the largest companies have even taken to calling their internal teaching program a university—perhaps something of a stretch.) In any case, the school metaphor does not have quite the focusing power of the war metaphor: in 6σ companies there are Green Belts, Black Belts, Master Black Belts, and Champions. Many of these positions are full-time, and the people who hold them are responsible for teaching and leading and for quality performance generally. All this assistance helps to remove the fear of the mathematical aspects.

As for the resistance to change, GE's Welch used plenty of carrots[16] and some big sticks. A CEO memo in 1997 told GE managers worldwide that further promotion depended on Green or Black Belt training; by 1999, the memo said, all of GE's professional employees (over 80,000 of them) must have begun 6σ training. When these edicts were placed in context with the standard GE policy of arranging, annually, for the lowest-decile performers to perform elsewhere, it is safe to assume that everyone whose attention could be gotten had it gotten.

Although this upgrading process did cost a great deal, the payoffs were even greater. The chief charges associated with poor quality, as would be obvious to an accountant, are inspection, scrap and rework, and warranties and concessions (see also Fig 5.9).

All this avoidable and wasted effort caused by low quality has been termed the Hidden Factory—a factory that costs a great deal and has no benefit. The 6σ philosophy is to shut down the Hidden Factory. Less obvious to an accountant, but still of great importance, are the more hidden costs of low quality, including the lack of customer satisfaction and the loss of market share.

5.3 LIFETIME, CHANNELS, AND NEWNESS

Before closing out the discussion of general marketing concepts, there are a few final product issues of which engineering students

---

[16]Lest anyone feel above this fray, Welch also tied 40% of his Vice Presidents' annual bonuses (i.e., short-term incentives; see §10.5) to progress toward 6σ goals.

should be aware. These include the idea of a product's lifetime, choice of the channels of distribution, and a closer look at the several meanings for "new."

Product Lifetime

New products are launched every day with much fanfare by their proud parents—the companies that developed them. Given the very small percentage of original product ideas that actually make it to market, this fanfare is well deserved. Still, even if the product is successful, it has a finite lifetime (Fig 5.10).

Figure 5.10: Product Lifetime.

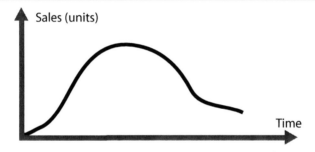

This curve is typical of the *semi*-quantitative graphs very common in business discussions. It is not representative of *any* mathematical equation; it has no great precision; it is merely a way of visualizing a concept and putting in a few rough numbers. Products are born; they live for a while; then they disappear, usually replaced with something better. That's all the graph says, plus information on how long the product lasted and its peak sales. After the fact, the curve is known via real data. But that is history and easy. The hard part, as usual, is to *predict* the sales behavior and thereby to execute the marketing strategy.

Channels Of Distribution

At the beginning of §5.1 we said that the three main marketing questions should be these: **(a)** What is my market? **(b)** What is my product? **(c)** How do I reach my market with my product? We have talked to some extent about the first two; now we look at the third question.

Figure 5.11: Channels of Distribution (examples).

Producer ➡ Customer

Producer ➡ Retailer ➡ Customer

Producer ➡ Wholesaler ➡ Retailer ➡ Customer

Producer ➡ Distributor ➡ Customer

Producer ➡ Agent ➡ Retailer ➡ Customer

Producer ➡ Agent ➡ Wholesaler ➡ Retailer ➡ Customer

Figure 5.11 shows various channels that may be used for various products in various businesses. For example, the large distribution of a widespread commodity may utilize the producer → wholesaler → retailer → customer channel. As a second example, consider the purchase of computer software.[17] Some companies sell it directly through downloads. Others use more complicated channels. If you bought MS-Windows®, it may have been preinstalled in a new personal computer you bought (one channel) or you may have purchased it as a box of CDs at a computer retail store (another channel).

When the flow of products and services is *outward* from one company to many customers, one refers to *distribution channels*; when the flow of products and services is *inward* from many companies to one's own company, one refers to the *supply chain*.[18] Both employ essentially the same concepts, and both require careful management, especially for large firms. The key point about distribution channels is that every step in the chain bleeds off some margin. Therefore, only steps that add some sizable efficiency boost to the channel should be included. The guy who says, "I can get it for you wholesale," is almost certainly getting in on the action himself. (It isn't really wholesale.) Moreover, if your company employs more than one distribu-

---

[17]And for those who pirate (steal) software, consider the purchase of software.
[18]These could just as easily have been called "distribution chains" and "supply channels."

tion channel, there will be an ongoing balancing act to ensure that the risks and rewards of the alternate channels are set fairly.

How New Is "New"?

The word *new*, as with so many words, has many shades of meaning. This is true even in everyday life. In business, anything that materially changes the business model has claim to some newness. For example, suppose a recently hired engineer finds a new manufacturing twist that saves 20% on production cost. The product itself is identical; only the cost structure is new. Does this qualify as "new"? Probably not to the customer—unless some of the savings are passed along to the customer in order to increase sales and gain market share. But to the business, this is an important innovation (Chapter 9) and there is a very welcome newness to the more efficiently manufactured product.

A study[19] that is a bit dated but still illustrates the point is shown in Fig 5.12. Six categories of "new" are used in a 2 x 2 plot of internal newness vs. external newness. For example, 7% of products are new in the sense that they are "repositionings." Here we have another example of new: same product, made the same way, but with new marketing parameters.

Figure 5.12: Sort-of-New, New, and Totally New.

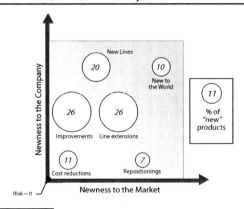

[19]*New Products Management for the 1980s*, from Booz, Allen & Hamilton, 1982, as quoted by OC Walker et al., *Marketing Strategy*, Irwin, 1992.

We also learn that, in this example at least, only 10% of new products are completely new products, meaning that they are new to everybody everywhere. There is an important lesson here for young engineers. The academic culture, driven by applied science values, suggests that most acclaim be given to completely new discoveries. This is consistent with the R&D responsibility for faculty staff. However, engineers can contribute in many more ways than creating products out of whole cloth. Students should know this if they are to keep all guns blazing in achieving their career goals.

5.4 MARKETING OF HIGH-TECHNOLOGY PRODUCTS

Virtually everything said thus far is true about marketing in general, although there have been many instances along the way where an enterprising young engineer with a broad range of skills can make a career-enhancing contribution. We now turn to those special issues that are related to the marketing of products that are so-called high-technology (or high-tech, for short). Perhaps we should pause first and explain what is meant by "technology."

The Many Faces Of Technology

Technology is a word whose popularity has exploded in recent years. It is therefore used constantly in our discussion, making its meaning one of the most important to clarify. A reasonable definition of "technology" is the following, from the *Oxford English Dictionary*:

> **Technology**
> The study, development, and application of devices, machines, and techniques for manufacturing and productive processes.

Perhaps the emphasis on manufacturing is a bit outdated. A more fundamental (and simpler) definition, created by the author for courses to young engineers, is the following:

Chapter 5: Marketing Concepts

> **Technology**
> Any man-made thing that one can use to assist one in doing something one wishes to do.

This definition is meant to include goods, services, products, and processes—and specifically also must include the mental component that facilitates the desired "use," namely, expertise, more colloquially called "know-how."

Figure 5.13: The Many Sources of Technology.

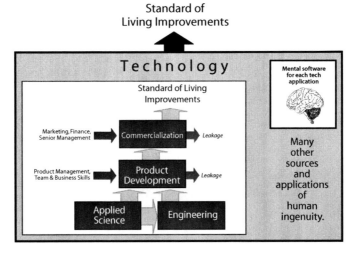

In fact, the more one thinks about technology, the more difficult the word is to pin down. Does it just mean *everything*? Not quite. But it certainly means more than just the output of a group of highly trained applied scientists, engineers, and business people creating products, services and processes. Although that description is apt for the subject of this book—and is shown in the block diagram in Fig 5.13—no definition of technology would be just or complete were it not to include all the other areas of human activity that also produce "something that one can use to assist one in doing something one wishes to do."

Here is an example the author hopes all young engineers can respond to.

> Antonio Stradivari (or Stradivarius, to use the Latinized version of his name) was born in 1644. He remained active in his shop in Cremona, Italy, until his death in 1737. His hand-constructed violins have been unsurpassed for more than 250 years. Stradivari also made harps, guitars, violas, and cellos—likely more than 1,100 instruments in all—of which ~650 still survive. Was he a mathematician? Nope. Was he an applied scientist? An audiologist, perhaps? It is to laugh. An engineer? Not at all. In fact, no one has ever found a satisfactory explanation for the transcendently beautiful sounds that come from the violins made by Stradivari and his Cremonese contemporaries—and it isn't for want of trying.
>
> Instrument makers have been attempting to reproduce every physical parameter they can think of, yet have failed to duplicate the magic. Physicists have used lab equipment to analyze the vibrational patterns of the key structural components of his violins, all to no avail. Chemists have cooked up elaborate recipes for the varnish that coats and colors a violin's raw maple and spruce, assuming it's the icing on the cake that counts. Again, no luck. Perhaps it's all a scam? Maybe it's been only mass hysteria for almost 300 years? Can these fiddles really be worth several $M each? While pondering this question, listen to Itzhak Perlman playing Shostakovich on his Strad, or any other of the top violinists playing a timeless classical masterpiece on the "products" of Stradivari, and the answer will occur to you.

Also shown in Fig 5.13 is an intangible that is crucial to the success of any technology—learning how to use it. The author dubs this the "mental software" associated with this technology, using the computer metaphor. (The best human computers are still more valuable and versatile than the silicon ones.) New tools require new skills. If one "wishes to do something" and uses technology to "assist" in that enterprise, one must learn how to use the tool. This can be relatively simple in the case of a potato peeler or quite difficult in the case of a baseball bat or violin.

Chapter 5: Marketing Concepts

There are also several additional useful phrases that incorporate the word *technology*, many of which are used and explained further as they arise in this book:

- A *new* technology simply means a technology that is of recent origin and that is still being absorbed by people and organizations. Obviously this is a relative term and, as time goes by, the new technology is supplanted by still newer technologies.

- *High* technology is one that the speaker or writer believes is especially difficult (to invent, to build, to use, etc.). This appellation also fades with time.

- A *product* technology is one that is primarily embedded in a product, service or device—the output of an organization's internal process. To civil engineers, the product may be a bridge; to electrical engineers, the product may be a computer chip; to chemical engineers, the product may be more efficiently brewed beer.

- A *process* technology is one that is primarily embedded in the organization's internal process itself. A wood-chipper is a process technology for a pulp and paper company.

- An *enabling* technology is one that permits entirely new process technologies to be developed and used cost effectively.

- A *sustaining* technology is one that an organization (or perhaps even an entire business sector) has adopted as foundational to its core competencies. Incremental innovations may continue but these can be further absorbed to advantage without paradigm shifts.

- In contrast to a sustaining technology, *a disruptive or nonlinear technology* is more than merely new. It cannot be incorporated within the current organizational framework, even though—because of its intrinsic advantages and perhaps its adoption already by competitors—it must be integrated soon, despite the great cost and temporary inconvenience. It's like the difference between evolution and revolution.

- A *strategic* technology is one that an organization sees as vital over an extended period of time.

- A *platform* technology is one that has many and diverse applications. It is difficult to find a better example than the modern computer.

- An *emerging* technology is one that is still on the horizon but that is likely to be widely disruptive. "Emerging" is sometimes used to capture simultaneously the attributes of very new, platform, strategic and disruptive. An example would be nanotechnology.

In the investing mania of the late 1900s, high-tech on Wall Street meant information and computer technologies.[20] That was a gross oversimplification, based on what was hot at that time. There are many other examples of high-tech, including biotechnology, advanced materials, pharmaceuticals, modern chemistry, high-tech energy, automation and robotics—the list is a long one.

Product Technology vs. Process Technology
One distinction related to the word technology that is worth making is *product* technology vis-à-vis *process* technology. The former connotes *what* is made; the latter *how* it is made. For example, a company that manufactures top-flight paper-making machines must find ways to create a machine that accepts porridge-like pulp, then convey it along a screen-based belt at speeds approaching 60 km/

---

[20] In fact, just before the bubble burst, Information Technology was just called "tech," as though there were no other technologies.

Chapter 5: Marketing Concepts

hr, while removing excess liquid and providing the nicely baked arrangement of micro-fibers of which we are unconscious when we read a newspaper or use a fine-grade paper. All this must be done reliably, shift after shift, ideally without a paper break or a stoppage of any kind[21] until the scheduled run is over. To that company, its paper-making machine is a product; but to its customers—companies that make paper—this machine is part of its process.

Figure 5.14: The Product-Process Chain.

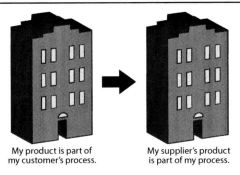

My product is part of my customer's process.   My supplier's product is part of my process.

This is not an isolated occurrence; on the contrary, it is the norm. As this is being written (a process), Microsoft software (a product) controls Intel chips (a product) in a Touch personal computer (a product). Three companies (and actually many more, of course) have contributed their products to the writing process (Fig 5.14).

This also raises the question: What, exactly, is a *high-tech* company? The usual mental picture is of a firm that produces products that can do miracles.[22] However, this places too much emphasis on the product half of the product-process linkage. Suppose a company that produces lawnmowers uses such high technology (for manufacturing and all its other processes) that it can sell its mow-

---

[21] Anyone who has seen the chaos that ensues when a paper break occurs on a high-speed paper-making machine will never forget the experience. All the paper any one of us will ever use or recycle in our lifetime comes flying out in a matter of seconds—a whirlwind that can be very costly if high-quality design is absent.
[22] As Sir Arthur Clarke is widely reported to have said, "Technology sufficiently developed is indistinguishable from magic."

ers at $10 and make a profit. (Of course, it will sell its mowers at a much higher price.) Would this company be classed as a high-tech company? One should think so! To judge a company's technology level, one must know not only the technology of what is produced but also the technology level of everything the company does. We shall revisit the product-process distinction later, in the chapter on innovation (Chapter 9).

Moore's Segmentation

The important process of segmentation was explained as a basic marketing concept in §5.1. The idea is to separate one's total potential market into submarkets based on one or more discriminating characteristics. This permits market targeting, which surely makes good sense. It would be rare to see an advertisement for a BMW in the *National Enquirer*. Usual segmentation characteristics for the general public include age, sex, lifestyle, etc.

Figure 5.15: History of First-Time Adopters of New High-Tech.

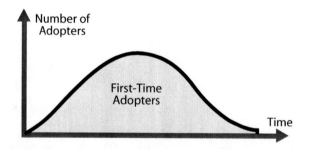

Geoffrey Moore wrote a series of books[23] on the marketing of technology-intensive products and developed a segmentation strategy based on *high-tech receptivity* (Fig 5.15). Like many good ideas, it seems obvious once conceived and explained. Although his primary scenario involved sales to other businesses, there is also much wisdom with respect to sales to the general population as well. The

---

[23]GA Moore, *Crossing the Chasm: Marketing and Selling High-Tech Products to Mainstream Customers*, HarperBusiness, 1991.

Chapter 5: Marketing Concepts

premise was not completely new—salespeople had noticed before that some customers were much more receptive than others to considering new high-tech products; they spoke of the *lead user,* a customer that enjoyed getting new technology for its new benefits, but Moore developed the concept much more completely. One does not need to use optimality theory and the calculus of variations to realize that the best, least risky move is to start with the most receptive customers, then move on to the next most receptive, and so on.

Figure 5.16 sketches most of the key ideas, but some further commentary should be helpful. The *Innovators* just love technology and like to try new things. They are not likely in top management (who would not like the high risk) but they are senior enough to be able to choose to be involved with a completely new product.[24] They are willing to help, make suggestions, and even β-test the product on their site, but they think of themselves more as a technical partner than a pure customer—especially when it comes to paying. Nonetheless, these are crucial receptors: they assist in many ways, and not just technically. They create credibility and referenceability for the product. There are very few of these folks around and they take some finding.

Figure 5.16: Moore's Segmentation for High-Tech Products.

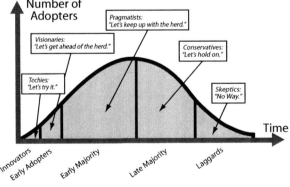

---

[24]Remember, we are talking about that minority of products that are in the northeast corner in Fig 5.12—the disruptive ones.

135

The next Moore segment, the *Early Adopters*—who together with the Innovators comprise the Early Market—are not so much technophiles as entrepreneurs (see Chapter 12). These folks don't just like fun new stuff; they really can envision the potential business benefits. They want badly to get ahead of their competitors and they can't do that unless they move fast. They are willing to go part way toward their supplier in creating interfaces and putting up with occasional glitches, but they can sometimes be demanding as well.

The Chasm

As Fig 5.16 shows, if there are no disasters in satisfying the first two segments, there are now large sales in prospect, as the *Early Majority* get exercised about your products. These people are neither technophiles nor visionaries; they are smart, pragmatic businesspeople. They know that change is the norm and they want to win from it, not be bludgeoned by it. Their due diligence on the experiences of the Innovators and Early Adopters leads them to believe that the risks are no longer forbiddingly high, but that the rewards are attractively so.

This is a critical point in the marketing of this new product, an epoch that Moore refers to as "the chasm" (Fig 5.17). A total change in corporate culture is needed with respect to selling this disruptive-technology product.

The marketing department should now spend all its energies selling to customers who basically want to buy; the engineering department should now focus, not on fooling around with minor improvements, and not on responding to fractious customers who want all manner of costly extras, but on building and shipping product (as designed) as fast as possible. The great moment everyone has been working towards is finally *here*; it would be a great shame were all the key players to be so focused on secondary minutiae that the team is not able to walk through this great open door of opportunity.

Chapter 5: Marketing Concepts

Figure 5.17: Moore's Segmentation for High-Tech Products.

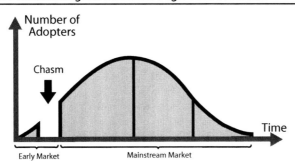

The different mindsets required for innovation vis-à-vis implementation are not only *different*, they are in many respects *opposites*, as argued by Martin and Austin.[25] Ambiguity must be replaced by predictability; faith must be supplanted by certainty; passion by discipline; and creative enthusiasm by business understanding. This massive cultural rearrangement often fails to occur, as does the planned financial growth (Fig 5.18).

Moore's final two market segments include the *Late Majority*, who by now have finally seen that they have no choice but to get on board, and the *Laggards*, who despite all their resistance to change finally sign the purchase order. The Laggards, frankly, don't matter much; they are too few too late. But the members of the Late Majority are many, and they extend the product's lifetime and help to provide cash for the next company innovation.

Figure 5.18: Many Do Not Successfully Cross the Chasm.

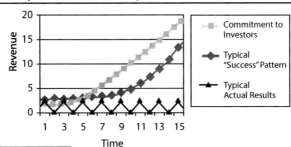

[25] RL Martin, H Austin, "Innovation vs. Implementation," *Rotman Management*, magazine from the Rotman School of Management, University of Toronto, Spring/Summer 2002, pp 6–11.

High-tech people in high-tech companies often get asked, not just about *selling* high-tech products *to* others, but about *buying* high-tech products *from* others. Even though, on the selling side, one loves an Innovator, being part of the Early Majority is probably quite soon enough on the buying side; this is probably the minimal-risk strategy.

CHAPTER 6

Politics at Work

Active Participants and Passive Victims

Author's Note: For this chapter, I have the pleasure of introducing a co-author—Neeraj Ghai—who has both a BASc in Mechanical Engineering and an MBA, and who is moving up rapidly within a major international corporation in the financial sector. Ghai's most important contribution was in convincing me that the subject of this chapter was critically important for engineering students considering their careers. I quickly agreed. If I had read this chapter as an engineering student 45 years ago, it would have changed my life. But office politics has always been a weak suit of mine, so I despaired of writing a strong chapter on the subject. Ghai responded by writing a draft of §§6.1–6.3. His strong performance as both an engineering student and an MBA candidate, and his impressive career success in the years since, give him solid credibility for this task.

Young engineers entering the business world can be forgiven for thinking that their educational attainments, impressive work performance, and admirable personal qualities will be the sole determinants of their success. Since this has been a successful formula in their careers thus far—and it is already a stretch to cite "admirable personal qualities" as essential (or even needed) in engineering school—it is understandable that they all presume that these same qualities, alone, will enable them to rise steadily up the corporate hierarchy and perhaps eventually land them in the coveted throne behind an executive's desk.

The fact is that education, hard work and performance are rarely enough. The higher one climbs up the management ladder, the more one realizes that, in the complex corporate world, what we shall call politics becomes a governing influence over one's rate of progression, career success and eventual level of accomplishment. Whatever one thinks of the desirability of this state of affairs, to be in denial is simply wishful thinking. How well one deals with power and politics largely defines whether or not one's career will incline upward, stagnate or even degrade.

6.1 CORPORATE POLITICS: AN INESCAPABLE REALITY

It is difficult to overstate the importance of organizational politics. It is often the determining factor in one's career and it is the purpose of this chapter to bring this fact home to those, especially engineering students, who have lived (or even flourished) in a culture where such considerations would be greatly frowned upon. These political smarts are learned more often the hard way than studied rationally and reflectively. Certainly the author has never heard of any course discussion of this topic in engineering school, during either his seven years as an engineering (applied science) student or his forty years as an engineering professor.

> **Definition of Politics**
> The art or science of governing.

The *Oxford English Dictionary* defines "politics" as shown in the accompanying box. There is no hint here that politics is an inherently bad thing, nor should there be, for how can one avoid some sort of process for resolving conflicting goals, principles or viewpoints within a group of individuals, whether the group is large or small, whether a whole country or just an organization?

When people, as they often do, profess a disdain for politics, surely it must be for the results of the political processes that affect them, or for the individuals involved that have political power, or for the basic political structure that governs their lives. Surely no thinking person could challenge the assertion that some sort of framework for resolving competing purposes must exist, and that, while such a framework is by definition a competitive situation, one cannot rail against politics just because it attempts to resolve the clashes between people and ideas.

Politics Is Competitive By Nature
Engineers, young or old, should not think that they are immune from the political processes in the workplace. Some may become so fed up with "politics" that they resign and start a small business of their own. A subset of these may be true entrepreneurs (Chapter 12).[1] Growing a new business does not avoid politics, although this arrangement does make it quite plain who is ultimately responsible for any corporate political mistakes. Other definitions of politics, such as "social relations involving authority or power," or "the often internally conflicting relationships among people," only make the point more strongly.

An important point to remember is that, whichever career ladder one hopes to climb (lifestyle businesses aside), this same ladder is also being climbed by many others. Among these others, some will prefer to leverage political strategy over performance and work ethic in hopes of boosting their careers. At all levels of management there will be people buzzing around, strategizing, manipulating and playing political games in hopes of arriving at their corporate destination before you do, even if it means stinging you in the process. Keep in mind that politics is often seen as a zero-sum game,

---

[1] On the other hand, those who start new, one-person "lifestyle" businesses may eliminate most of the challenges of workplace politics in their lives, but there are many other irritants to take their place. This career choice may be the best for some, although it is more difficult for engineers than for some other professions.

which means that in order for someone to win, someone else must lose. In the course of their climb to victory, some colleagues will not hesitate to step on you.

A Turn-on? Or A Turn-off?

It is immediately clear why it is important that all students be aware of these realities. Some, whose gifts in the areas prized in engineering school (mathematics and science) are rather limited, may feel that the political arena is where they will shine. They would then tend to take optional courses that prepare them better for management—or even a program in engineering management if they are fortunate enough to be enrolled in a school that provides such a program (see Fig 3.1). But they must know about office politics in the first place.

Contrariwise, there may be students who are attracted by the greater authority and higher pay levels associated with management, and who plan to enter this field at some future time, but who are repulsed by the attending political realities. These students may then work even harder at their engineering (applied science) courses, or even take advanced applied science degrees, knowing that they are doing so for the long haul. But again, they can't make such early career guidance corrections if they are unaware of corporate political realities.

Chapter coauthor Neeraj Ghai makes another solid, related point:

> Some readers may be thinking to themselves that corporate politics is not their thing and that they do not want to involve themselves in such games. But even if one refuses to play, declining to step on others will not protect one from being stepped on. Refusing to sting others will not protect from being stung. Those who are politically savvy will continue to impinge on your job satisfaction, organizational commitment, morale, working relationships and conflicts, performance appraisals, promotional opportunities and salary increases. In the worst of cases, the dynamics of politics can result

in severe job dissatisfaction, potential leave-taking, or termination. Your career may even enter paralysis.

Likewise, you have the opportunity to protect yourself (often at the expense of your rivals or allies) if you are an active participant in this game—and at the very least, maintain a sense of political awareness. In short, political behavior in organizations (a necessary evil for engineers wishing to enter the world of management) and your ability to engage in it can either do you wonders or leave you in dazed wonderment. Make no mistake; either you are an active participant or you will become a passive victim.

Welcome to the reality of corporate politics.

Purpose Of This Chapter
Engineers wishing to become managers need to accept and be aware of corporate politics, reframe their negative perceptions of it, and become more politically savvy so that they can seize opportunities as they arise and protect themselves from the pitfalls of blame, failure and difficult power struggles.

The purpose of this chapter is not to instruct the reader on how to become politically savvy; that could be the topic of an entire book on its own. Instead, our intent is to **(a)** highlight the significant role of corporate politics in management; **(b)** identify the main types of political strategies and tactics used by individuals in the workplace; and **(c)** provide enough insight into corporate politics to enable the reader to assess his or her ability to navigate the political environment. Our underlying premise remains this: Whether or not a reader begins to realize that entering management is the right idea for him or her, an essential first step is to consider the realities of corporate politics.

6.2 UNDERSTANDING POLITICS
Although some companies are more political than others (see §6.4), politics has at least some effect (perhaps a major effect) on one's ability to achieve one's career aspirations and performance goals. An important

start along a march toward success in management is to understand the difference between corporate policies and corporate politics.

Politics vis-à-vis Policies

Corporate *policies* refer to documented rules and standard operating procedures that encompass the methodologies and constraints for getting the job done. Policies educate new employees about the basics of the job and serve to provide useful, detailed information on what work needs to be done and how one should go about completing it.[2]

For our purposes, corporate *politics* can best be described as a game of power, where one attempts to influence key decision-makers through the use of discretionary behaviors whose primary purpose is to promote personal objectives. Politics can reflect three types of personal agendas: **(a)** pursuit of position, **(b)** control of resources, and **(c)** advancement of individual goals.

What makes corporate politics difficult to understand and gain leverage from is the fact that the tactics needed and used are (unlike policies) undocumented. These tactics are not prohibited by anyone or any company policy. Instead, they are unwritten rules of how things are done or not done. Not surprisingly, understanding these unwritten rules may differentiate one favorably with respect to one's coworkers who do not, bringing benefits during performance appraisals, allocation of high profile assignments, and when promotions or layoffs are being considered.

Neeraj Ghai contributes the following examples from his personal experience:

> I once knew a lady who worked for a family business for over four years. Over the years she was praised for her work, given increasing responsibility and enjoyed good raises. Everything was great until a new fellow—a member of the owner's family—came into her department. Once that happened, her boss (a VP), a non-fam-

---

[2]Those whose duties are purely *administrative* spend all their time applying corporate policies. They are not permitted any wiggle room for exercise of intelligent judgment in particular cases not adequately covered by the policies.

ily member, began delegating to this relative all the responsibilities she had once handled. In addition, her boss began spending two-hour lunches with the relative. She approached the VP about this and other situations that had been taking place. His initial response was very defensive and two weeks later she was fired. Naturally, she felt wrongfully discharged and felt she could not go back into an office environment where she might again fall victim to office politics.

In another instance, I have a friend who is a human resources (HR) generalist, making about $40K per annum. Recently, she told me that her career goal is Senior VP of HR (where she would make over $100K) but does not think she'll ever make it because she does not like to play corporate politics. Many of my technically oriented friends often tell me that they are happy with their jobs and are glad they do not have to put up with corporate politics.

Before considering a career in management, one should first accept politics as an organizational reality. Even though some companies are less political than others, none can claim to be free of it.

The Good, The Bad And The Ugly

An important first step in coming to terms with the reality of politics is to examine—and, if possible, redefine—its negative perceptions. Certainly the family-business example cited just above is on the ugly side, and there are many other stories of the same ilk. Politics is often viewed as the nasty side of the corporate game because it has been credited with affecting diversity in corporate America, especially at the power levels. Indeed, to some, the word "politics" used in connection with corporate culture primarily implies backstabbing. Most people believe in merit; they do not like backstabbing, taking credit for others' work, or succeeding on personality rather than performance. The ugly side of politics usually involves unethical behavior; ethical persons, including engineers and managers, do not behave unethically. Indeed, the most successful careers in the long term require a high level of personal integrity.

Thus we see that not all politics should be viewed as bad. Politics also includes the art of communication. To a great degree, corporate politics means developing good *people skills*, and there is nothing ugly about those.[3] At its best, politics involves contributing more than expected, being diplomatic, collaborating and cooperating, and conducting a low-key public campaign for oneself.

Politics Reframed

Successful managers think about corporate politics as being about relationship building, networking and gaining visibility so that they can better influence people, sell their ideas and get into positive power positions. This enables them to benefit their company (and thus themselves). Those that are politically knowledgeable will be able to network effectively, build immunity from negative politics by promoting their own contributions, and build relationships with powerful seniors to increase their influence. To be really successful, one must not only reframe corporate politics, but also understand the critical role of communication within it. Typically, top executives have mastered communication: *overtly*, by being open, direct, and honest; and *covertly*, by developing personal rapport behind the scenes.

Figure 6.1: Reframing Corporate Politics.

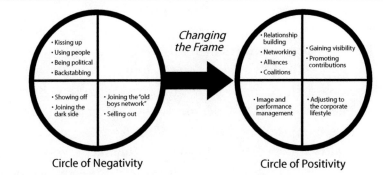

---

[3]Employers of newly minted engineers are unanimous that the skills most conspicuous by their absence are people skills, also called *soft skills*. This should not be surprising of new engineers, given their education. More surprising, perhaps, is that the employers of new MBA graduates have precisely the same complaint, although they are presumably using a higher scale of expectations.

Chapter 6: Politics at Work

Successful managers realize that creating relationships with people who work for them and with them is as important as interacting with superiors. This awareness prompts managers to learn how to influence others on a personal level to generate results. We must reframe corporate politics, understand it, embrace it and leverage it to our benefit and for our protection (Fig 6.1).

Just as one person sees stubbornness, another, more positive person will see persistence. Fig 6.1 shows how to view more positively many of the actions and attitudes of other workplace individuals.

The Best Strategy

To move up the corporate ladder, one must be a political player. This means improving communications, learning to compromise, making others (especially your boss) look good, being assertive (without being aggressive or abrasive), building effective networks, taking on leadership roles, developing friendships (not a growing list of enemies) and continuously placing the corporate mission and agenda ahead of your own. In order to help the reader develop this kind of strategy, the six most common tactical concepts in corporate politics are presented in the following section.

6.3 SIX COMMON TACTICS FOR WINNING AT CORPORATE POLITICS

To help develop a better appreciation for the political dimension of management, this section presents an overview of six common tactics used by managers:

> **Six Common Tactics Used by Managers**
> 1. External Attribution of Blame
> 2. Controlling Information
> 3. Credit Appropriation
> 4. Building Networks and Coalitions
> 5. Crafting Obligations
> 6. Spinning Images

More than one of these (indeed, all six) can be used and will interact, as shown in Fig 6.2. Some of these tactics are unethical or can be used either ethically or unethically. Nevertheless, they will all be described since an understanding of all is mandatory if one is to mount appropriate defenses. Even if one is personally ethical (the best long-term strategy), one must still fend off the unethical machinations of others.

Figure 6.2: Tactical Considerations for Corporate Politics.

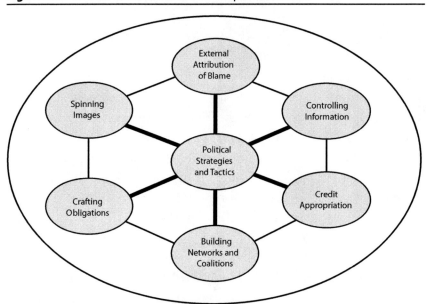

Tactic 1—External Attribution Of Blame

The most direct form of organizational politics in which some people engage is blame shifting—part of what some call the "blame game." This tactic refers to assigning responsibility for negative outcomes in the workplace to rivals (including former allies). These people attempt to frame their failure in terms of external attribution; for example, they may try to explain the tardiness of an overdue report on the lack of support from another functional unit or on other conditions beyond their control. This happens every day in companies large and small, and at all management levels. People will say and do anything to save face. And it isn't always so overt. It also includes the subtle tactic of dissociating oneself from undesirable situations.

Why are such dubious games played? Because there are many notable benefits to inaccurate attribution. First, blaming others helps minimize embarrassment, guilt, remorse and anxiety over a botched project or bad results. In addition, it may help maintain one's status, rank or reputation amongst the seniors in the face of a potentially problematic situation, because it assigns ownership over failures, shortcomings, and mistakes elsewhere, thereby avoiding punishment. Often, it also allows one to provide a simpler, more acceptable explanation for why things are the way they are. The avoidance of accurate attribution also prevents empowering rival individuals and avoids having to grapple with complexity, ambiguity and responsibility.

Blaming others, however, does not come without serious risks. These include: a diminished sense of one's own integrity, a reduced ability to remedy weaknesses (often people choose not to fix what they choose not to acknowledge) and a reputation for being political, unfair, or agenda driven. In the eyes of the politically savvy, these tactics are rather obvious and it may be you whom they victimize. The blame game often results in personnel turnover, diminished performance, reduced commitment and morale issues for peers and subordinates.

The fact is that most individuals naturally want to take credit for their successes and blame external causes (including others) for their failures.[4] Is blaming others worth the loss of trust, respect and loyalty from one's coworkers? Threatened by potential job discontinuation or tempted by imminent promotion, the answer *may* be yes. However, it should be clearly realized that much of Tactic 1 involves lying, which is unethical.

---

[4]Some say that much of society has migrated to the stance that nearly everyone (meaning everyone who is not highly successful and productive) is a victim, meaning that no one is responsible for his or her own behavior and state. If this is true, then it is not surprising that individuals who are surrounded by fervent believers in victimology at home, school, and within their peer group join the workforce thinking that it is normal to blame others for their woes.

Tactic 2—Controlling Information

Information is political. Controlling information can be source of power or defense. Careers can be significantly impacted based on how information is shared. In every company, there are those who will control information and increase their own status by strategically controlling how and with whom information is shared. Likewise, there will be those who protect information and keep it closely guarded so as to get a leg up or to make themselves appear indispensable.

Once again, chapter coauthor Ghai has a personal experience that makes an excellent example:

> In my relatively short career, I have seen some engage in negative politics by organizing meeting agendas to suit their personal schemes. In one case, two competing employees were going to attend the same meeting: Employee A was responsible for the success of Project X, while Employee B was responsible for the success of Project Y. Employee A set the agenda for the meeting and placed issues relating to Project Y near the bottom so that the attendees would be too fatigued to be interested in making careful judgments—or would not get to the Project Y issues at all.
>
> The following week, Employee A arranged a meeting to make shared resource allocations at a time when Employee B was on vacation. When Employee B returned, he was behind the eight-ball in terms of both knowledge and available resources. To make matters worse, A's and B's performances were being measured on equivalent scorecards relative to one another. A truly zero-sum game!

There is also the classic case of a boss and his ambitious, high-performing employee. To put the matter bluntly: If your (more senior) manager feels that you may be making him look bad—because you are a superstar employee that is increasingly being recognized and credited for successful results from your department—your manager may change the way he interacts with you. Projects that would once have been assigned to you may be assigned elsewhere.

Likewise, information that you were once given may no longer be provided. The point should be obvious: If your own boss turns off the knowledge tap, you will with time lose any potential to succeed. By controlling information, your boss or rival employees can limit or even destroy your opportunities.

Another Ghai example:

> Several months ago, I found myself in a situation at work where one of the senior executives asked me to audit a particular process. The operations manager in that area seemed somewhat paranoid and became so concerned that I would uncover something that did not reflect well on her that she contacted her direct reports and asked them not to provide certain information to me. Yet, without this information, I could not fulfill the requirements of my task. I felt strangled because this operations manager was using information control as a defensive mechanism, perhaps so that her incompetence would not be revealed through my process improvement efforts. She cost me the opportunity to deliver on a high-profile project, while saving herself from being exposed—all through the basic tactic of information control.

Now that so many members of the workforce are characterized as knowledge workers (Chapter 10)—and certainly all engineers fall into that category—the opportunities for (and consequences from) impeding the proper flow of information are increasing every day. Once again, many of these information control tactics are unethical, but that doesn't stop the ethically challenged from using them.

Tactic 3—Credit Appropriation

We begin with a typical scenario:

> Donna was finally fed up and though she was not looking forward to complaining to her boss, David, she realized that things had reached the point where she had little choice.
>
> Donna began, "Steve lied to Jennifer about who does the forecasting of call volume and now every time Jennifer asks him for

more details, he stalls and tells her, 'Leave it with me, and I'll get back to you.' Then he privately asks me for the information. It makes me want to puke. I do all the work and Steve takes all the credit!"

"And you were telling me that something else just happened, right?" David inquired supportively.

Donna replied, "Yes, I just found out that Steve didn't understand Jennifer's modification request. So the alterations I made to the forecasting model aren't even remotely what she wanted. Now it's too late to make the appropriate changes to the assumptions. Worst of all, Steve is now trying to blame me and is telling Jennifer that it's my fault. This is sickening."

In every company, there will be managers and supervisors who take all the credit for the work of subordinates or others who are less powerful or more vulnerable. This devious political tactic is called credit appropriation, and is characterized by the appropriator taking credit for the work of the targeted individual, who is more often than not a subordinate.

To prevent getting entangled in the consequences of credit appropriation, one should be aware of patterns of appropriation and of who the players are that use this tactic. By maintaining a sense of awareness, one can be more prepared if one is targeted by the appropriator. In addition, it is important to remember that, if appropriated work falls short of expectations or involves some sort of glitch, stolen credit may be returned in the form of attributed blame. In order to protect oneself against *credit appropriation*, one may wish to seed one's work with complexities and nuances that only they understand.

By leveraging and controlling this cryptic information (see Tactic 2 above) one can sometimes protect oneself against credit appropriation. As a form of theft, Tactic 3 is also unethical.

Tactic 4—Building Networks And Coalitions
Another political tactic that managers must learn to master is building strong networks with key players. This involves cultivating

trusted colleagues and even social relationships for the purpose of fulfilling personal objectives. By leveraging networks, politically aware managers can get valuable information before others do, which assists them in making better and more timely decisions. Through networking they can also gain informal approval from allies and sponsors—which makes it easier, later, to gain formal approval for projects and other initiatives. Finally, networking provides an earlier chance at perceived job opportunities, while the job description is still fluid and the level and pay range can still be influenced.

Having one's own professional network has become important to the point of being indispensable in this era of ever-more-common mergers, acquisitions, downsizing and restructuring. In today's high-performance environment, one is continuously in peril from downturns in the economy, changes in one's home industry, corporate politics (need we add here) or just plain bad luck. It is our networks that can often protect us. If one fails to build and maintain an active network, one not only forgoes the opportunity to proactively exploit it, one exposes oneself to the slings and arrows of outrageous fortune[5] by not having its protection against the professional maneuverings of others.

An added difficulty for women in particular is that they are often excluded from male networks. Since men remain the dominant power holders in most organizations, this poses problems for women who wish to join the ranks of senior management. Fortunately, these barriers are slowly being torn down.

Organizations themselves engage in networking when they need new employees. If one seeks a new or better management position, the aim should be to make sure that their networks intersect with yours. Networking will generate many opportunities and personal referrals, enhancing your success rate at identifying available management positions as well as your bargaining position with respect to them.

---

[5]With apologies to Hamlet.

Building stronger, long-term relationships that have future potential when it comes to job search and getting information or resources should take more than the 0.0001% of your time that your training as an engineer might suggest. It is important to let others know what your current activities are, meet them for lunch occasionally, stay in touch with classmates and colleagues and take leadership positions in both your professional and personal life. Meet diverse groups of people and increase your visibility, in turn increasing your power to tap them for information and further contacts. And pass the vision along: motivate others to broaden their networks (they will return the favor one day). If you wish to succeed in a management career you must have exceptional networking skills.

One or more of the *dramatis personae* in your network should be a mentor. A *mentor* within your company can help you understand its culture and how to navigate around politics—and possibly to help you be considered for plum assignments. No matter how limited their time, most successful people are genuinely interested in helping others succeed; they benefit from the experience also. If you are truly set on being a manager in the business world, you must feel comfortable approaching people, asking them to be your mentor and building an enduring relationship with that person. Mentors can be anyone in your field (broadly defined) and should have more industry experience under their belts than you do, even if it is in a different industry. Mentors can refer you to jobs, to managers who are hiring, and can act as references on your behalf.

When entering the business world, one may also find oneself exposed to, or involved in, a coalition. A *coalition* is a group of people who elect to demonstrate their strength in numbers by pooling their resources and power in hopes of influencing key decision-makers to make decisions in support of the group's common objective. Coalitions often demonstrate a sense of urgency and broad support because of their size and scope. These groups are better able to command support than efforts made by individuals pressing for the same cause.

Tactic 5—Crafting Obligations

We again start to understand a new tactic through a parable from Neeraj Ghai:

> A while back, I met with a friend of mine (Susan) who was quite happy with herself. She told of how she and her colleague (John) were working on a project together.
>
> During a conversation between Susan and John's boss (Peter), Susan told Peter many wonderful things about John and his performance on the project. Since Peter did not work in the same city as his direct report, and since these comments were made close to performance appraisal time, they significantly skewed the performance rating that John received from his boss (in a positive manner). Susan met with John and when John was telling Susan about his performance rating, it became apparent to both of them that Peter had basically used Susan's comments verbatim to prepare John's performance assessment.
>
> Susan chuckled and took this opportunity to remind John that he owed her one. A few weeks later, Susan's performance rating was being discussed around the table. In attendance were both Peter and John. Someone else at the table questioned Susan's rating, implying that it should be lowered. However, John voiced his strong positive opinion of Susan and helped her maintain her performance rating. John did this because, in part, he felt obliged to. He "owed her one."

The obligation to help someone who has helped you is a natural consequence of being helped. Many individuals use this as a political tactic for support and promotion. For example, an employee who seeks your help on a particular project may then feel indebted to you. You may wish to lever this feeling of indebtedness by asking that person to put in a good word for you to your boss, provide support for your ideas, or help you make a lateral move in your company by using that person's network.

People are more likely to help and support you if they are made to feel obligated. Crafting such obligations can be a key political tactic in getting what and where you want; the politically astute are sometimes able to reap rewards greater than the original debt owed. Of course, one must be careful not to upset anyone, destroy relationships or burn bridges.

Tactic 6—Spinning Images
Communication skills are mandatory in demonstrating a high level of professionalism. One must be able to communicate well with everyone above, below and beside. Managers should understand that a low level of communication skill could prevent them from climbing the ladder of success. Remember that professionalism also includes how others perceive your behavior.

Successful managers are successful spinners of images. In particular, they are able to put across an image of themselves that is viewed as desirable by their firm for someone in their position. In order to be successful in management, one must try to understand the image (both in terms of appearance and behavior) that upper management desires—and then cultivate that image. The first step is to understand your corporate culture and company environment. By understanding important corporate elements such as communication style, teamwork, chain of command, appearance, decision-makers, management roles, interoffice friendships, politics and individual attitudes, one can better acclimatize oneself to one's environment. Such an understanding will impact day-to-day decision-making, including choices pertaining to dress, language and behavior, and will be especially important during face-to-face time with seniors or clients.

A senior leader at a large Canadian media company explained, "We have a dress code at this company that allows people to wear jeans.[6] But it's important to remember that one

---

[6]When making important presentations, Stephen Jobs wears the oldest, most faded jeans he can find, complete with gaping fissures in the cloth. Although this display of abject poverty seems rather phony for one who was a multi-millionaire before age 30, it has for some reason become almost *de rigueur* in the computer business. Anyone who imitates this sort of behavior does so at their own risk.

should dress for the job they want, not the job they have." Managers who understand the importance of spinning images and strategically capitalizing on both the appearance and the behavior prized by their company culture are far better off when climbing the corporate ladder. In some types of business, image is everything. It's essential to have senior management believe that you are competent, professional, intelligent, reliable, productive and deserving. Creating such an image may even protect you if you ever make a mistake or botch up a project—or become the victim of another political player engaging in the tactic of external attribution of blame.

The author has a recollection of his own:

> Several years ago, I had a meeting with one of our company's consultant lawyers at one of Toronto's leading accounting firms. (Yes, he was both a lawyer and an accountant!) It was a very warm, muggy, August day. After offering the mandatory coffee and juice, he invited me to remove my suit jacket. Being used to the disheveled dress code now in vogue for engineering professors (and many high-tech managers), I was more than willing to accept his offer. I noticed he did not follow suit (pun intended) and asked about it.
>
> "You certainly won't offend me by taking off your jacket," I said. "I know you're just as good a lawyer when you're comfortable."
>
> Thanks for the sympathy," he said, "but our dress code requires traditional business attire whenever we're with a client."
>
> As I was leaving, he invited me to join him that night at a Raptors NBA basketball game. I enthusiastically accepted. As I slid into the seat beside him that evening, I noticed that he was still wearing his business suit—here, where everyone seemed to be competing with the team mascot for the strangest dress style. Somewhere in the middle of our first beer, I gingerly inquired, "I know you are still technically 'with a client,' but does your firm want you to wear office garb even to a basketball game?"

"I really don't know," he smiled, "but I have enough legal conundrums to face every day without worrying about that one."

He had, of course, also made another point—that he took his job *in every detail* very seriously. That sense of commitment, presumably, included the quality of his advice to his clients.

If the image you have created for yourself has been intelligently spun, a mistake will be viewed as being uncharacteristic and will not likely lead to any sort of repercussion. However, if you are viewed as being incompetent, a major blunder may cost you your career. If you cannot manage and shape the impressions others have of you, others may spin those impressions for you. So before you enter the business world with high aspirations, ask yourself how strong you are at managing, shaping and spinning impressions. They will be all-important in your climb up (or tumble down) the corporate ladder.

6.4 FINDING YOUR POLITICAL FIT

Some engineers may feel that the political dimension we have been discussing gives them an edge over their more bookish colleagues; they may naturally seek a position in an organization where these skills result in career benefits. Others may be distinctly uncomfortable in a highly charged atmosphere and feel that such shenanigans are not what they are about, not why they went to engineering school, and not what they wish to think about at work. Still, the discussion thus far in this chapter has made it amply plain that it is impossible to avoid politics entirely (§6.1), and that politics can often be a positive thing (§6.2).

Degree Of Politicization

A mature approach to the subject is not to shun politics altogether (naive), nor to make it the all-in-all of one's daily activity (superficial). At these extremes lies failure. The objective should be to find an appropriate fit between one's own political intensity and that of the organization one works for. A detailed, rational approach to finding this fit has been given by Reardon, whose work[7] is the basis for Fig 6.3.

One begins by quantifying the degree of politicization of an organization. Reardon's framework is shown in Fig 6.3, which has four levels (it could have been three or five) ranging from virtually zero to pathologically intense—the last group being literally sick from politics.

Figure 6.3: Degrees of Organizational Politicization (based on Reardon).

Minimally Politicized	Moderately Politicized	Highly Politicized	Pathologically Politicized
• Amicable atmosphere • Conflicts rare, brief • Rules bent to grant favors	• Rules agreed and understood • Focus is on growth, agility • Capable of positive change	• Conflict pervasive • Rules known, but often flouted • In-group vs. Out-group • Upward communication weak • "Who" more important than "what"	• Headed for self-destruction • Conflict pervasive • Massaging of information • Productivity feeble • Growth stunted • Subordinates unappreciated

Which Degree Of Politicization Is Preferable?

Reardon does not overtly express an opinion about which level of politicization is best; she is interested primarily in raising awareness and helping readers find the best working environment for themselves. The present author senses that the "minimally" politicized organizations, although they are likely the most comfortable environments, would not produce good career results—especially if the organization were a small company (meaning that it must fight for its revenue and growth). Careers are best at companies that are growing, and growth cannot happen without some process to resolve the inevitable conflicts that arise.

Really capable people know they are really capable; they are rarely satisfied by being ignored as such. They expect career re-

---

[7]Kathleen Kelley Reardon is a professor of management and organization at the USC Business School. See, for example, her article "Managing Internal Politics" in *Business: The Ultimate Resource*, Perseus Publishing, 2002; or her most recent book, *It's All Politics: Winning In A World Where Hard Work And Talent Aren't Enough*, Currency Publishers, 2005.

wards but such rewards (other than lip service and counterfeit promotions) are unlikely to happen at the quiet, pastoral, static, "minimally" politicized company. It might be inferred that an unpoliticized staff may be the most comfortable, but not likely the most dynamic or effective. If a young engineer wants to just curl up in a corner and read a book (say, on Bessel functions) and not interact much with colleagues, the lack of politics is a boon. However, if that young engineer wants a productive level of excitement, a frequent and meaningful interaction with other team members, and satisfying career growth, he or she should seek out a company that has healthy politics to manage growth, change and conflict.

It seems equally obvious that the "pathologically" politicized company has become fanatical and has lost its purpose, its mission, and its mind. Everyone is focused on themselves and the team synergies are all negative. Reardon gives four of the symptoms that enable one to diagnose this syndrome:

- People in power are overly flattered by subordinates, while people in weaker positions are abused.

- No one wants to rock the boat and the primary means of communication are lying, hints and innuendo.

- Malicious gossip and backstabbing are ubiquitous.

- No one is valued and everyone is dispensable.

It is difficult to imagine why any capable person would want to work for such a stressful, depressing circus—especially since its long-term business prospects are poor. No career there either, unless one is a relatively incapable (not to mention unethical) person who can thrive only through meanness and street smarts.

That leaves the moderately and highly politicized companies; in the author's opinion, preference is strongly dependent on personal taste.

6.5 IN CLOSING—TWO THOUGHTS

What we have been calling "politics" is a reality that must be deftly handled. Perhaps everyone should have the aphorism in the accompanying box on their desks.

> **It's your** *ATTITUDE*,
> **not your** *APTITUDE*,
> **that determines your** *ALTITUDE*.

The main message of this chapter is reflected in this quote from Reardon:

> Why do some people who work hard and effectively at their jobs fall behind, while those who are adept at "reading the office tea leaves" forge ahead? Being politically savvy doesn't necessarily mean being unethical or devious. At heart, it's about listening to and relating to others, and making choices that advance everyone's goals. Like it or not, when it comes to work, it's all politics. And politics is all about knowing what to say, when to say it, and whom to say it to.

Reardon is speaking to all prospective employees. This book has a narrower audience: engineers who are interested in careers in management. For them, abstaining from politics is not an option.

A last thought in closing is this warning: You are always on display whenever you are with other people. Your colleagues will remember, many years later, that time at the holiday party when you behaved inappropriately to a junior employee (or any of the myriad other opportunities to make conspicuous mistakes). Even at lunch or at so-called "off site" events, and in similar contexts, don't try to pretend that your work associates are temporarily your bosom buddies. They are still your associates and one wrong move will be remembered—and when recalled much later will overwhelm a hundred fine accomplishments. If you want to be a senior manager or organizational leader some time in the future, start thinking and acting like one very early and very consistently.

PART II

For Young Professional Engineers

Can You Evaluate Your Management Opportunities?

Engineering school is now behind you. You learned a great deal in school, but were likely surprised at the many differences between the demands of a realistic engineering position and the visions conjured up in undergraduate engineering class. While you are excited by the reality of now holding a well-paying position in the profession you decided many years ago to pursue, you also have become aware of important differences between the realities of your job and your earlier expectations.

One of those differences, certainly, will be your new perceptions of *management*. You begin to realize that the organization for whom you work has goals that may not coincide with yours. You are being asked to perform functions that were not stressed in school—and may never have been mentioned. You begin to realize that you have assumed a role in a team game, and you must either recognize and embrace that fact or work for many years as a disappointed—possibly bitter—person.

You may decide to respond to management requests professionally, playing the role for which you are paid. To be successful in this quest, you must realize that everything in the real world, including your own activities, are fraught with risk. Knowing the sources of these risks (and how to cope with them) is a key survival skill (Chapter 7).

Chapter 8 faces squarely an organizational activity that you may have thought was either above you or below you: accounting. As one aspires to be an influence at ever higher levels in one's organization, one must learn in ever more detail the advantages and disadvantages of the established methods of keeping track of its financial success.

The concept of innovation is attractive to the engineer, especially one who seeks to have his work lead to more than papers or reports. Chapter 9 demonstrates that innovation implies a broad team effort and that a young engineer who decides to lead this process should guide himself into engineering management. The generalization of this process is wrapped up in the concept of Intellectual Capital (Chapter 10).

CHAPTER 7

Some Risks Are Worth Taking

Thorns and Roses along the Path

It may seem unconventional to place a discussion of *risk* at this juncture, so a brief justification may be in order. We could treat risk as a *qualitative* entity, a word useful to accompany handwaving and rhetoric. Certainly the word is quite familiar and most have a view on how it should be used. However, we believe that risk should be elevated to the status of a *quantitative* idea, one of the inroads that mathematical modeling can make into business strategy discussions.

Given their mathematics courses, young engineers are as familiar as they will ever be[1] with the subjects of statistics and probability. These mathematical tools comprise the vital bridge between a more primitive world (where risk is just a matter of opinion) and the developing world of management science (where risk is an irreplaceable modeling tool for making smarter business decisions).

This validation for taking the subject of risk seriously segues nicely into the second reason for our prominent discussion of risk: many business scholars and experienced managers have come to believe that much of management is an exercise in *risk management*. This percep-

---

[1]Unless they plan to study these subjects further through majoring in applied science research.

tion will only intensify and clarify as risk analysis grows from being based mainly on gut feel to becoming a body of quantitative analysis.

We shall briefly discuss the risk-reward relationship at the end of this chapter (§7.6). When businesses and their managers contemplate moving into new areas of commercialization, they must evaluate accurately the risks and the rewards (the costs and the benefits) of doing so. Both must be calculated with comparable accuracy if the best decisions are to be made.

The familiar word *"game"* is helpful in this context. Commercial activity (meaning legitimate businesses run by ethical management) can be viewed quite accurately as a *game*—where no disparagement or trivialization is intended. To be something other than a childish diversion, a game must have some degree of complexity; it must permit the continuing exercise of intelligence; and it must be played against competitors who have some accomplishment in the *genre*. And—most important for the present discussion—a game must have uncertainties.

The classical board games all have elements of uncertainty. No serious book on bridge or backgammon fails to have chapters on probabilities, and even chess, that pinnacle of logical board games, requires that each player make judicious assumptions about the other player's intents, experience and personality—clearly with probabilistic components. Thus, the nugget of the issue is *trying to predict the future*. Of the various professions considered in §3.5, only those involved in business and commercialization attempted to predict the future.

7.1 WHAT IS RISK?

How does one predict the future? In many scientific contexts, predictions can be made to a high degree of accuracy. That is why those people who are guided by reason give scientific results so much respect (§4.3). One can plan for years to land a command module on the moon and then do so (though certainly not with zero risk). Unfortunately, most situations in everyday life are fraught with uncertainty and business is no exception. There is no complete "science of business" and none is in imminent prospect (to put the matter gently).

Those who claim to help people with predicting the future—from the seers and prophets of old, to the stock analysts and "psychics" of our present time—have always been predictably able to make a decent living, but business managers are well paid only if most of their predictions are *correct* and this surely makes them more deserving of respect. Thus, no discussion of business strategy—and especially no discussion of the commercialization of radical innovations—can be complete if it does not recognize, either implicitly or explicitly, the inherent risks.

Risk analysis starts with the recognition that no one can predict the future with complete accuracy. It continues by seeking to identify (and to find relevant information about) all the trapdoors over which the planned path may pass. And it finishes with a complete analysis, based on mathematical principles wherever possible, of the chances of each trapdoor opening.

Risk Requires A Probabilistic Approach

In evaluating a business opportunity, or in deciding on the wisdom of developing a new product or market, sometimes the outcome of a certain event will be binary, meaning that the outcome must be one of two values. In effect, the outcome is either 0 or 1. Either it happens or it doesn't. For example, a potential first customer may say they will let you know by the end of the month whether they will purchase the first unit or not. One must then estimate the chances of 0 vs. the chances of 1. In many other situations, a large number of possible outcomes are possible and one must estimate the distribution of probabilities around these outcomes. Although this procedure does seem somewhat complex, even a semi-quantitative treatment is better than no treatment at all. Some have defined a great mind (and others have defined a great education) as one where complexity and ambiguity are tolerated. If there is one human activity that requires tolerance for complexity and ambiguity, that activity is leading-edge business.

When one first learns about probability theory, the primitive example tends to be the *coin toss*. This simple scenario is undeniably sufficient to make several key points about probabilistic theory, among them the following: **(a)** some events—all events, as it turns out—cannot be reliably predicted, hence we need a means for dealing with this uncertainty; **(b)** since each event in such a chain is independent of all previous events in the chain, runs of bad (or good) luck are statistically expected; **(c)** even though all the detailed aspects of a coin-flip are governed by the laws of mechanics, aerodynamics[2] and structures—established beyond any real doubt for over a century—the application of these laws even to the seemingly simple situation of a coin-flip are so crushingly complex that a more effective modeling approach is to pretend that all is determined by the Probability Gremlin.

We should pause and remark that this last resort to probabilistics is employed in many complex situations, both business and non-business. In modern physics, many processes are believed to be inherently probabilistic (just mentioning the name Heisenberg should suffice); there is no deterministic alternative. But other fertile grounds for probabilistic models are situations with intractable deterministic models. In most business risk calculations, any improvement in technique would be a boon. A wild guess on high vs. low is not enough—although, at the other extreme, massive applied science models, even if available, that spend precious resources on the next decimal place are rarely justified.

To return now to the key lessons that can be learned from even a simple coin toss, we also discover **(d)** the criticality of assumptions, such as the fair coin assumption, which assumes that heads and tails are equal each toss; and **(e)** even if the coin is not fair, one can collect statistics on the heads/tails ratio for that particular unfair

---

[2]In case any reader is following this argument in minute detail, we should also note that the effects of air turbulence are themselves treated in a probabilistic manner.

coin, so that, based on these statistics, one can modify one's probability distribution (H vs. T) for this coin. This last point establishes the essential link between the pedestrian world of merely counting occurrences, to the focused activity of assembling statistical models that will, in turn, be used to build a theory for the application with which one is concerned.

An Example Of Mathematical Probability

Simple estimates of probability, as called for by risk analyses, are well understood in mathematics. Fig 7.1 shows a simple probability distribution, called the χ^2 distribution,[3] but the exact shape of the distribution is unimportant. Here, we simply use it to think about the (unknown) probability of some business outcome. For example, the abscissa could be the number of units sold per week. On the top is shown the distribution function (meaning that the probability that we sell between 5 and 10 units next week is the area under the curve between x = 5 and x = 10). On the bottom is shown the cumulative probability, so that the chance of selling at least 5 units is the ordinate of the curve at x = 5. Clearly, the cumulative function is the integral of the distribution function, and the distribution function is the derivative of the cumulative function.

Figure 7.1: The Chi-Squared (χ^2) Distribution: (a) The probability density function; (b) The cumulative probability function.

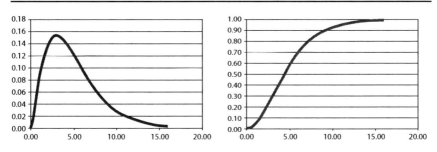

---

[3] The particular one shown has five "degrees of freedom" and is often used by experimenters to test "goodness of fit," although these specifics are not relevant to our present discussion.

We may not be sure in advance how many units will be sold, but if we adopt the model of Fig 7.1, this is equivalent to assuming that there is only a 5% chance that our average weekly sales will be less than 1.15 units, and there is a 95% chance that we'll sell less than 11.07 units per week. (This is exactly equivalent to saying that there is a 95% chance that we'll sell more than 1.15 units per week, and a 5% chance that we'll sell more than 11.07 units per week.)

Incidentally, the expected sales rate (the *mean*) in Fig 7.1(a) is five units/week; the *mode* (the peak of the curve) is three units/week; and the *median* (the 50-50 point) is 4.35 units/week. If our cost analysis has shown that our breakeven sales rate (the number of units that will permit us to pay all expenses, but have no profit left over) is four units per week, then we're in a pretty risky business: We have just seen that, although we expect to sell five units/week, there is equal likelihood that we will sell *less* than 4.35 or *more* than 4.35.

This is a good example of *semi-quantitative* analysis, which is often useful in business decisions. Unlike many situations in *science* (flipping coins and drawing cards are the simplest examples) we do not know the *a priori* probability distribution for the critical variables we are considering—and this is the typical environment for business decisions. In the sales projection above, the exact situation has never been seen, evaluated, experimented with, and measured to provide the beginnings of a probability distribution like Fig 7.1.

This means that the analysis is less precise than it might first appear. In reality, it is just informed, intelligent guesswork. *Semi-quantitative* analysis is not as comforting as *quantitative* analysis, of course; but if there is no solid basis (math model) for truly quantitative analysis, then semi-quantitative analysis is better than hand-waving or crystal-ball gazing. The best people simply make the best estimates they can. We learn that, if we can get our costs down a bit further, such that the breakeven sales rate is three units per week, then this business proposition should probably go ahead (unless there is an even better use of capital); it will be risky, but the upside reward (also based on probability) considerably outweighs the downside risk.

Chapter 7: Some Risks Are Worth Taking

The two skills most needed for doing these semi-quantitative analyses are not mathematical techniques, but experience and judgment. Often, the decision-maker finds even a semi-quantitative probability distribution like Fig 7.1 more than can be guessed at, and he or she decides on the expected value (a.k.a. mean value, or average value).

7.2 HUMAN RESPONSES TO RISK

Before getting into some mathematical and practical details of our approach to taming risky threats, two fundamental points must be made that will condition our discussion.

The First Human Response To Risk: Aversion

The first is a fact about (most) human beings (see box) and this observation is germane because it is, after all, humans who populate a business's customers, suppliers, employees and partners—and its investors, shareholders and managers.

> **The First Principle of Business Risk**
> Most people are averse to risk

To illustrate this principle, consider the following mental experiment:

> At the end of every workweek in a certain mining community, the workers are paid their wages as they pass from the mine for a well-deserved weekend's rest. Normally, they simply collect their paychecks, but on our hypothetical occasion the paymaster makes each of them an interesting proposition. "If you choose to flip this coin," the paymaster says, "we will tear up your check if you flip tails, but the company will pay you double if you toss heads. Do you wish to participate in this game?"
>
> One of the younger miners, who was once in engineering school, says, "I've had some training in probability theory, and I happen to know that, if your coin is 'honest,' which I believe it is, my return

on the flip will be zero 50% of the time and double 50% of the time, meaning that my expected gain from this game, over and above my normal paycheck, is zero. But I like a little excitement, so I'll flip the coin, knowing that my expected gain (and loss) is nil."

Most of the miners, however, are older. They have wives (or husbands) and children and mortgages. They understand from their educated co-worker that they can't really win anything in the long run from this sort of bet, although they can have a little fun trying. Their second thought is that, although the probabilities are symmetric (equal likelihood of +1 paycheck and –1 paycheck, depending on the flip) the consequences of playing this little game are far from symmetric. If the flip turns heads, they and their dependents will have twice as much money to spend for the coming week: perhaps for that desired article of clothing; maybe for that necessary improvement to the home; or a contribution to that well-earned vacation. On the other hand, if the flip produces a tail, they will not be able to provide even the necessities of life to their loved ones during the coming week: neither food, nor clothing, nor habitation, much less the frills.

Most of the miners decline the offer to flip the coin.

Thus we see the basis for the human aversion to risk. The pain of the downside is always greater than the pleasure of an equal upside. Also note that both the upside reward and the downside risk must be taken into account. This aversion to risk is not confined to miners who are living week-to-week and paycheck-to-paycheck. Even wealthy investors, if asked whether they would derive more happiness from an upstroke of 10% or misery from a downstroke of 10% during the same period, would prefer not to play at all.

Before continuing, let us agree to ignore those who are simply gamblers. These folks are not just indifferent to risk (as with the mineworkers in the above example, who chose to bet their paycheck with no expectation of net reward); they welcome it, even when all their experience indicates that they are losers. Intelligent business risk-taking, however, is not gambling.

Businesses that try to avoid all risk by remaining stationary become paralyzed by the accumulated environmental changes that affect them, and they succumb to the long-term risk of being left behind, perhaps fatally so, by these inevitable changes. If, in the end, an enterprise loses from taking considered risks, it can at least be consoled by the knowledge that it played the game as best it could, rather than lose for certain by not playing.

Aversion To Change

> **Corollary to The First Principle of Business Risk**
> Most people are averse to change.
>
> **Niccolò Machiavelli on Change:**
> "There is nothing more difficult to carry out, nor more doubtful of success, nor more dangerous to handle, than to initiate a new order of things."
> —from *The Prince*, written over five centuries ago.

A corollary to the principle of risk aversion is the aversion to change. If today's actions are the same as (or very similar to) yesterday's actions, which are also last week's actions, then increasing comfort grows with the status quo; the uncertainty of the surprises that come with change, this person reasons, must be avoided. When change finally comes, the person's environment and the attendant risks become more unnerving. This process manifests itself as an aversion to change.

Retuning one's responses to the changing environment takes more time and effort, until the new learning curves are traversed, and this leads to psychological stress. Of course, there are a (very) few who actively welcome change, some to the point where they are almost flighty. At the other extreme, some suffer from such a powerful aversion to change that they slowly cease to function effectively as the organization changes around them; these individuals should remember that not to change is not to learn, and that "no decision" (i.e., postponement of a decision) is itself a decision.

The Second Human Response To Risk: Underestimation

It is interesting that the second tendency that humans have toward risk evaluation tends to compensate for the first. Perhaps matters are balanced after all. In any case, let us return to the mental experiment of the mineworkers above, and note that, at least for them, the risk (probability) analysis was starkly simple, with a ±1 outcome. Their response to the coin-flip challenge was not a matter of estimating probabilities; it was a matter of psychology and priorities.

Not so in the real world of business. Most of this chapter is devoted to the evaluation of risk for business propositions, particularly for those that are based on new technologies, or for those based on new products that use older technologies. In the semi-chaos of reality, there are no simple theories of probability to rely upon. Even worse, the data are incomplete, possibly unreliable, and likely out of date (Fig 3.5 has outlined the situation).

> **The Second Principle of Business Risk**
> Most people underestimate risks.

Once again, we emphasize that we are talking about predicting the future, and that this undertaking is impossible to perform with the accuracy we should desire. It is also a psychological reality that a stance of pessimism rarely produces impressive results. The object of the game is to gain a reward, while risking a loss. The positive attitude of energy and enterprise that is so conducive to making good things happen is also inimical to being overly cautious. Thus, even though businesspeople are personally averse to risk, they often compensate by underestimating the actual risks involved.

This tendency usually takes one of two forms: **(a)** the risk in each particular link in the chain of outcomes is underestimated; and **(b)** the number of links in the risk chain is underestimated. Fig 7.2 shows a simple such chain, for which

$$p(F|A) = p(F| \cdots) \cdots p(\cdots |C)p(C|B)p(B|A)$$

Chapter 7: Some Risks Are Worth Taking

In words, for a three-event chain, the probability that Event C will happen (given that Event A has happened) is equal to the probability that Event C will happen (given that Event B has happened) x the probability that Event B will happen (given that Event A has happened). The Long Path in Fig 4.1 is another example of a long, complex probability chain, with the "leakages" corresponding to times when the events do not contribute to activities farther along the path.

Figure 7.2: A Simple Probability Chain.

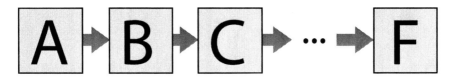

All probabilities, of course, satisfy

$$0 \leq p \leq 1$$

with $p = 0$ corresponding to certainty that the event *won't* happen, and $p = 1$ corresponding to certainty that the event *will* happen.

Regarding the probability of events happening, we commend the following corollary for reflection:

> **Corollary to The Second Principle of Business Risk**
> The only way that one can be certain that a particular event will happen (i.e., the only way that there is zero risk of its not happening) is if it has already happened.
>
> **Hughes's Addendum:** Even then, don't be too sure.

In addition to the psychological mechanism underpinning this principle (optimism with respect to the "known unknowns," as predicted by the Second Principle of Business Risk Calculation) there is also the phenomenon[4] of the "unknown unknowns." One must also be careful that the unknown unknowns are not merely ignored unknowns, where discretion fails and valor may have to struggle on its lesser parts.

Finally, lest any reader think that it is just a feeble attempt at humor, Hughes's Addendum has a serious intent. More pointedly, how, exactly, can one be certain that an event has, in fact, happened? How many times have we all laughed nervously at bromides such as "The check's in the mail," which is a parable illustrating this same corollary. Even more subtle: If the check has been received in the mail, and opened for verification, and deposited in the bank, has it cleared the system? (Note that even this simple example has a chain of four identifiable events.) Anyone with experience has many examples of people phoning to say that they have decided to do something, but the something never actually happens.

The box below contains a simple exercise that will be educational to those unaccustomed to pondering risk.

Exercise for the Reader

Step 1: Think of some event you are planning, or would like to see happen, such as a vacation, and construct all the sub-events that must happen in order for a successful vacation to happen. **[The Event Chain]**

Step 2: Identify, in detail, all the reasons why you would like this event to happen. **[The Reward]**

Step 3: Estimate the chances of each sub-event happening. **[The Risks and Associated Probabilities]**

Step 4: Finally, based on your responses to Steps 1–3, what do you think the chances are of this event happening? Is the reward large enough to justify all the effort and cost? **[The Business Decision]**

---

[4]*Known unknowns* refer to mechanisms that are known to have some effect, but whose precise influence is not known, even in the probabilistic sense. *Unknown unknowns* are the surprises later that were not considered, even superficially. Careful thought can often place some of the "unknown unknowns" into the "known unknown" category, thus enabling sharper risk evaluation and improved decision-making.

Chapter 7: Some Risks Are Worth Taking

Simple Mathematical Insights

The phrase "risk chain," as in Fig 7.2, has been chosen because of the old (and true) adage that a chain is as strong as its weakest link. Thus, if most of the probabilities are high, but one is low, that is bad news. One doesn't average the probabilities in the chain; one multiplies them! For example, suppose one has a business decision and careful analysis shows that, from beginning to end, there are 100 separate things (links) that have to go right. Suppose further that 99 of these 100 links have a very high chance of success (say, $p = 0.98$) and that one unavoidable link has the discomfiting chance of one in four (0.25) of achievement. Then the overall probability of success is

$$p = (0.98)^{99}(0.25) = 0.0338$$

that is, 3.38 per cent.

From this, we learn two things: **(a)** a deal that might loosely be thought of as a "98% deal" can be brought down quickly to failure by a single low-likelihood link in a long chain, and **(b)** since clearly the single 25% link can't be blamed for bringing the "98% sure thing" down to less than one chance in 25, we also learn (with apologies[5] to Everett Dirksen) that "a percent here, a percent there, and pretty soon you're talking about a big loser."

In fact, closer examination of the above calculation shows that

$$p = (0.98)^{99}(0.25) = (0.135)(0.25) = 0.0338$$

So we see that the 99 links that are virtually sure things (each 98% certain) together create twice the problem that the one weak link does. In combination, these two effects make the deal hopeless.

This same effect is sometimes couched as the following riddle:

---

[5]Everett Dirksen (1896–1969) is alleged to have said in the U.S. Congress, "A billion here, a billion there, and pretty soon you're talking real money."

> You are mingling in a very large crowd of people. With each person you meet you strike up a brief conversation in which you include the question: "What is your birthday?" The riddle for you is this: How many people do you expect to have asked before someone says that their birthday is the same as your own?
>
> Most people intuitively answer, "365," based on the assumption that babies are born more-or-less uniformly throughout the 365-day year.
>
> We shall use the same assumption about the uniform distribution of births, but we shall get a different answer. The probability that the first person asked has a different birthday is $p = 364/365$. The probability that the first two people asked have a different birthday is $p = (364/365)^2$, and so on. After 253 people have been asked, the probability that they have all given a birthday different to yours is $(364/365)^{253} = 0.4995$. In other words it is more likely than not that someone in the first 253 asked will have the same birthday as yours. This is, once again, a demonstration that many events, each very nearly certain, can mount up to a perhaps surprising degree.

A second example is important because it illustrates that one doesn't make all decisions based solely on odds:

> An even simpler (and more lethal) example is Russian roulette, wherein one chamber (only) of a six-shooter is loaded with a live round, and then a series of "spin the cylinder and pull the trigger" moves is made. The chances of being alive after one event is $p = 5/6$. After two, it is $p = (5/6)^2$, and so on. You start to lose an even-money bet if you bet on a group of three trigger pulls. Perhaps, then, you should make an even-money bet after each shot? The odds certainly favor it. You will win five times in six.

If, by losing, you lose not only your wager, but much, much more, then you should decline even a "winning" bet. This has many applications in business—for example, one's policy on taking out insurance policies. Since the odds are always in the insurance company's favor—they have to pay all expenses, build big buildings, and show a profit out of the difference between premiums paid in and claims paid out—the odds are always against the policyholder. One should not buy insurance unless either **(a)** losing without insurance would be a catastrophe, or **(b)** someone else is paying the premiums. Insuring one's life may make sense; insuring one's bicycle may not, even if the odds of collecting are identical.

7.3 FIRST COMPONENT OF BUSINESS RISK: TECHNOLOGY RISK

With the above concepts in hand, we now move on now to consider the calculation of business risk. This has to be general enough to be useful for a wide variety of situations but not so general that business really doesn't come into it. The discussion will focus on certain key risk categories, after which the total risk is built from applying the principles of §§7.1 and 7.2.

Technology Risk

All engineers should become familiar with the notion of *technology risk*—how to calculate it and how to reduce it. Technology risk is connected to the following basic question about a new product: Will it work? Even the simplest questions can have hidden complexity, and in "Will it work?" one must ask what "work" means.

> **Technology Risk Component**
> Basic question: Will the new product work?

Here are some vignettes:

- The patent says it should work.
- Dwayne had it working the other day on his bench in the lab.
- It works at our strategic partner's location, although only Freda can get it started.
- It works at Customer X's place of business, but we still have to babysit it a lot.
- It worked well, and the customer liked it, but who knew it would wear out so fast?

As suggested by these comments, the only opinion that finally matters is the customer's opinion. The new product should work, reliably, and as advertised, and should continue to do so over the promised product lifetime.

If it doesn't work, using the above interpretation, all sorts of maintenance and corrective actions will be expected by the customer (perhaps costing many times the price of the product), costly redesigns will need to be launched, the goodwill (and willingness to act as reference sites) of the critical first customers will have been squandered, and the future of this commercialization venture will be bleak.

Technology Readiness

Evidently, the idea of technology risk is closely associated with *technology readiness*. Many schemes have been created that quantify the level of readiness, having many features in common. The Mankins[6] system of risk management uses no fewer than nine TRLs (Technology Readiness Levels) as shown in Fig 7.3. (A few words have been generalized from the aerospace context to the more general business context.)

Figure 7.3: Technology Readiness Levels for Risk Management (Mankins)

	DEFINITION
TRL 1	Basic principles observed and reported
TRL 2	Technology concept and/or application formulated
TRL 3	Analytical and experimental critical function and/or proof-of-concept
TRL 4	Component and/or breadboard validation in laboratory environment
TRL 5	Component and/or breadboard validation in relevant environment
TRL 6	System/subsystem model or prototype demonstration in a relevant environment
TRL 7	System prototype demonstration in final environment
TRL 8	Actual system completed and qualified through a rigorous test plan
TRL 9	Actual system proven through successful customer operations

---

[6] JC Mankins, *Technology Readiness Levels*, White Paper, NASA Advanced Concepts Office, 1995.

Chapter 7: Some Risks Are Worth Taking

To use the earlier terminology of Fig 7.2, the Mankins model for technology evolution (Fig 7.3) posits nine identifiable links in the development chain, each of which contributes an important risk source. Fig 7.3 should also be compared with the Long Path, the generic value chain for this book (Fig 4.1). There are many other similar algorithms for measuring whether the technology risk is being squeezed out of the product as planned in the budget and schedule.

The formula for evaluating the technology confidence is

$$p_T = p_{TRL1} p_{TRL2} p_{TRL3} \cdots p_{TRL8} p_{TRL9}$$

where p_T refers to *probability of technology success*, or *technology confidence*, which is the opposite of the probability of technology failure, or technology risk, here denoted by q_T:

$$p_T + q_T = 1$$

Thus a confidence of 1.0 means it is certain to happen; a confidence of 0.0 means it is certain not to happen.

This relationship also holds for each of the links in the process:

$$p_{TRLj} + q_{TRLj} = 1$$

for $j = 1, \ldots, 9$. Note that, while the overall confidence can be calculated by

$$p_T = \prod_{j=1}^{9} p_{TRLj}$$

the overall risk is *not* given by

$$q_T = \prod_{j=1}^{9} q_{TRLj}$$

but by

$$q_T = 1 - \prod_{j=1}^{9}(1 - q_{TRLj})$$

Examples Of The Technology Readiness Levels

Here are a few examples of the TRLs in Fig 7.3. If TRL-1 is an event that has happened, this means that the basic principles have been observed and reported. The scientific basis is solid, and thus, compared to the Long Path (Fig 4.1), TRL-1 corresponds roughly to the science being complete (no leakage, in this case). Similarly, completion of TRL-2 means roughly that the applied science has been completed. To complete TRL-3, one must have a design completed, a complete math model, and some experimental evidence that the concept will work.

At this point, we should remark on a new complexity. Products frequently have several components (some have hundreds). The risk analysis is carried out for each component until the point of integration, where the risk analysis continues for the product as a whole. Fig 7.4 provides a simplified illustration of this situation, with two components. Note that the probability of Event C happening is based on two probabilities, one for each component.

Figure 7.4: A Simple Probability Tree.

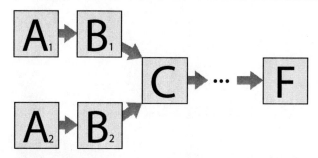

TRL-4 is often called the α-version, meaning a fully functioning version of the design on the company's home territory. It applies, as do TRL-1, TRL-2 and TRL-3, to each component. TRL-5 is often called the β-version of the component; it is no longer in the forgiving environment of its home laboratory, but it is still in a relatively friendly environment, perhaps in the field but attended by company personnel, or in the laboratory of a strategic partner or technology-friendly customer.

TRL-6 is a test of the various *components* as assembled and integrated into their *subsystem*; TRL-7 is a test of the various *subsystems* as assembled and integrated into the final *system*. Note that *two* levels of assembly are accommodated: components into a subsystem, and subsystems into a system. The generalization of Fig 7.4 should be obvious. If the final product (system) is simple enough that two levels of assembly and integration are not needed, levels TRL-6 and TRL-7 are merged.

TRL-8 indicates that the product must be put through a complete program of testing by its engineers and knowledgeable users in the relevant environment. This test program would include checking for a long list of possible failure modes. Every time a failure mode is proven not to exist, some more risk is squeezed out of the system.

Finally, TRL-9 *happens* when the customer has used the product in normal operations for an agreed period of time. Customers have a talent for finding problems that no one else can.

7.4 SECOND COMPONENT OF BUSINESS RISK: MARKET RISK

New technology is always risky, especially if the business has a standing army of researchers and engineers to support. Even if all these researchers, engineers, and applied scientists have performed brilliantly,[7] however, and produced a magnificent new product opportunity, there are many other major sources of risk. Not the least of these is market risk.

---

[7] Brilliant by business standards (meaning developing products that will surely be bought), not by research standards (meaning writing papers that will surely be published).

Market Risk

It should be plain from Chapter 5 that selling a new product, even if it has solid technical credentials, is a minefield. These adverse influences fall into a category called *market risk*. The most risky time for a new product is at the beginning. (It is difficult to avoid biological analogies.) Presumably this product, if it is being offered on the market at all, has successfully passed a number of internal hurdles.[8] It may not be absolutely perfect, but it is basically sound.

> **Market Risk Component**
> ***Basic question:*** Will they buy it?

Now the question is: Will they buy it? There are many ways to go off the tracks, as indicated by the following vignettes:

- They won't mind paying that much for something as good as this.
- I have three engineer friends and they all think it's neat.
- We haven't worked out yet exactly how we're going to sell this, but we're interviewing a salesperson next week.
- We're assuming our customers will make their purchase decision more-or-less right away.
- We haven't heard about any competitors yet.

Once again, as always, the customer is king. They buy what they want, when they want, and on the terms they want, including price.

Also, on the subject of competitors, even if there are no competitors in the sense that someone is selling substantially the same product as you are, there are always two intangible competitors—substitute products (as taught by microeconomics), and *just not buying*. This latter is the most obvious one, yet is often forgotten. Your customer has apparently lived a relatively comfortable life up to now without your product and so why do we assume he cannot

---
[8] See the discussion of project management in Chapter 9.

continue to live comfortably without it? Everyone's discretionary spending is greatly constrained, and every company's budget has a long list of "nice to haves." The substitutes go well beyond those referred to in microeconomics.

Market Readiness

Unfortunately, the author is unaware of any stepwise system for assessing market readiness in the style that the Mankins system (Fig 7.3) does for technology readiness. However, two of the basic questions asked at the beginning of Chapter 5 (What is my market? How do I reach my market with my product?) already give some structure to the process. Market research is essential before investing very much in a new product. This is a large subject in itself.

In some ways market research is easier than applied science and engineering research, but in others it is more difficult. Predicting the movements of the celestial orbs has at least the assurance that the governing laws have been the same for billions of years and will continue to be the same for billions more. In contrast, Aunt Martha's decision to buy something tomorrow may be quite different than it was today. People say things in focus groups that turn out to have a degree of unreliability about them. Segmentation, positioning and forecasting are not exact sciences either.

However, at some point, one is justified in saying, with a certain level of confidence, that the market has been identified. Completion of this process, which may well be going along in parallel with product development if the signs are good, eliminates a large amount of market risk. Then one turns to the second basic question: How do I reach my market? The prominent techniques for accomplishing this task have been sketched in §5.3, and as the channels are (first) chosen and (second) established, the chances for nasty surprises continue to diminish. The market risk has again been materially reduced.

Technology Risk And Market Risk Combined

In reality, one rarely solves the "market" problem completely and then swings over to developing a product that will fill an unmet market need. And anyone who completely develops a product with no knowledge of the market either suffers from OCD or is being paid by some organization that doesn't depend on the market for survival.

Fig 7.5 shows graphically how to think of the simultaneous development of product and market potential. The abscissa is the *market confidence*, p_M, which, as usual, bears a simple relationship to the *market risk*, q_M, thus:

$$p_M + q_M = 1$$

Similarly, the ordinate is the *technology confidence*, p_T, which is related to *technology risk*, q_T, thus:

$$p_T + q_T = 1$$

Every product, in the course of its development, is located at a point in this diagram. Four specific products are shown (as bubbles), and at the time the diagram was made, they vary in their technology and market confidences, as shown.

Figure 7.5: Market Confidence + Technology Confidence.

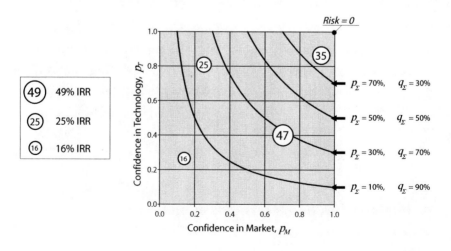

Chapter 7: Some Risks Are Worth Taking

The number inside the bubble is the IRR (internal rate of return),[9] an indication of the estimated profitability of the product when completed. This is like a third dimension to the plot and is an indicator of the size of reward expected to compensate for the risks taken. For example, the product that has the IRR of 35% and high confidence looks like a big winner. Even the product whose IRR is 47%, though currently at lower confidence (higher risk), is potentially so profitable that the risks should be taken. The product languishing at low confidence levels and an IRR of 16% should likely be abandoned.

Right at the origin, confidences are zero, and risks are both 1.0 (i.e., certain failure). Diagonally across, at the northeast corner, $p_M = p_T = 1$ and $q_M = q_T = 0$, indicating that confidences are complete and all risks have been driven to zero (an idealization). The confidence in the total product would be

$$p_\Sigma = p_M p_T = 1$$

and the total product risk would be

$$q_\Sigma = 1 - p_\Sigma = 1 - (1 - q_M)(1 - q_T) = 0$$

The curves in Fig 7.5 are curves of constant product confidence (and therefore of total product risk), as labeled.

Technology Push vs. Market Pull

Every product tends to start its life somewhere near the origin (where risks are high and confidences are low) and progresses to the northeast corner as product and market development continue. This assumes that the expected reward is justifying this development. If a company is be-

---

[9] If one calculates the net present value of all present and future cash flows relating to a product's business lifetime, including all flows out associated with costs to develop and market, etc., and all flows in associated with revenue expected—and if one uses IRR as the discount rate—one gets zero. This is the *definition* of IRR. The higher the IRR, the greater the reward, and hence the greater risk that can be justified in bringing the product forward. (See more on the risk-reward relationship in §7.6.)

ing run by engineers who have not learned to balance their passion for new technology with an equally strong emphasis on market work, they will tend to march rather vertically, developing the technology with an insufficient eye on the market. They may even believe that nasty old lie about the world swarming to their lab if they build a better mousetrap.

They may also be so enthused about their product that they have unrealistic expectations about how many customers are willing to pay how much money, causing an overestimate of its profitability. One cannot assess profitability without the necessary market research and development. This is an example of technology push, as exhibited in Fig 7.6.

Figure 7.6: Two Unbalanced Approaches.

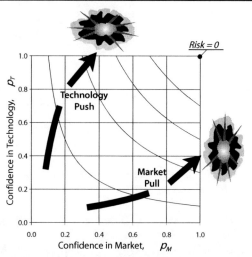

Technology push is far too risky a strategy (more likely, the absence of a strategy), because one can find out too late—that is, after too much has been spent on product development—that the market just isn't there. There may be no price point at which this product can profitably be sold: raising the price loses customers, and lowering the price loses margins.

However, there is a symmetrical trajectory that can also lead to foundering on the rocks of risk, also shown in Fig 7.6. Although market pull is almost always preferable to technology push, it its extreme form it can represent a group of gung-ho sales and marketing people promising customers a wonderful product—with features that may be far too difficult (expensive) to develop or manufacture. It the most extreme cases (and these do happen) the laws of physics would have to be violated to produce the pledged customer benefits.

Some of the best long-term business successes have combined market pull and technology push. Thirty years ago there was virtually no market pull for personal computers. People didn't know what they were or how to use them, nor could they afford them. So it would be hard to argue that personal computers were all market pull. However, as people learned what PCs could do they were willing to try them out and a strong market pull developed, both for the devices themselves and for all the software that gave them full potential. During the period of greatest sales, both push and pull were exerting full force, creating a Category-5 business hurricane.

7.5 THIRD COMPONENT OF BUSINESS RISK: MANAGEMENT RISK

Two major sources of risk—technology and marketing—have now been discussed, and we turn to a third category: *management risk*. This refers to the level of confidence one has in the management team to allocate the available financial resources and to guide the available people to a successful execution of a well-drafted business plan. Research has shown that technical people (engineers and scientists) underrate the importance of management, and place it well below technical skill, product design and market receptiveness on the risk importance scale. This is part of a larger principle that most people underestimate the importance, workload and skills associated with everyone above them on the organizational chart. The most general principle is the Principle of Vocophilia (see §3.4).

Contrariwise, venture capitalists, whose profession is to assess the overall likelihood of business success (at least for small, growing companies) and who put their money where there opinions are, rank management risk as either the single most important risk, or second behind market risk, depending on which one is asked.

> **Management Risk Component**
> **Basic question:** Do they know what they're doing?

So the question now is: Do they know what they're doing? Again, there are many ways to go off the tracks, as indicated by the following vignettes:

- There isn't a good marketing person on staff yet, but we plan on hiring one as soon as funds permit.

- Lynn seems to be surprisingly good at selling things, even though she's an engineer; she tries to contact potential customers when she's not busy with the design.

- The CEO just brought his brother in to do the accounting; he says his brother is quite good.

- I've never seen the plan and I don't think there is a written one; probably the CEO has it in his head.

- Nobody around here ever gets any feedback on how well or poorly they're doing.

Management isn't as easy as it looks, and it's clearly very important. There is no reason management positions can't be held successfully by people who were originally trained as engineers; still, like everything else worth accomplishing, it takes focus, natural ability and a great deal of hard work.

It is hoped that this book will help young engineers to recognize these career options. The author has been at many business conferences and has heard bankers, angel investors, marketing specialists, strategy gurus and others intimate that the worst thing anyone can do for a high-tech business is let the engineers run things. (Perhaps they haven't heard of Andy Grove, Bill Gates, David Packard, Jeffrey Skoll, or hundreds of others.) Still, they have a point. Excellence in engineering certainly does not, by itself, confer excellence in management, any more than *vice versa*. Great dentists, pharmacists, accountants, racecar drivers, physicians, farmers and airline pilots also won't be good managers unless they wish to become managers, have the talent to become managers, and make the multi-year commitment to become managers.

Combining Management (And Other) Risks
If one wishes to quantify the total risk, including management risk, q_L, one can calculate

$$q_\Sigma = 1 - p_\Sigma = 1 - p_M p_T p_L = 1 - (1-q_M)(1-q_T)(1-q_L)$$

One may wish further to include financial risk, q_F—for example, the effects of currency fluctuations, inflation rates changes, etc. Or perhaps even government risk, q_G—for example, unfavorable tax changes, export duties, or other changes in regulations that affect the business, one can calculate

$$q_\Sigma = 1 - p_\Sigma = 1 - p_M p_T p_L p_F p_G = 1 - (1-q_M)(1-q_T)(1-q_L)(1-q_F)(1-q_G)$$

These semi-quantitative kinds of calculations do not have great accuracy and would make a mathematician wince, but they produce better management than just hoping or praying.

7.6 THE RISK-REWARD RELATIONSHIP
In science, one is familiar with the existence of *laws*, where the word has a special (and thrilling) connotation. Not laws in the sense of

government statutes, or SEC regulations, or even parking rules, but laws that have a grandeur only good science can provide. The laws of motion and the law of gravitation, for example, promulgated by Isaac Newton in mid-age near the end of the 18th century, are to this day the basis[10] for understanding the motion response[11] to gravity-force *of all bodies in the universe,* from Olympic divers to the orbit of Titan around Saturn.

> **The Risk-Reward Principle (Weak Form 1)**
> Given two transaction opportunities, both with the same expected reward but with different expected risks, most humans will take the one with the lower expected risk.

In business, and more generally in management, there are unfortunately no such epochal laws. (And there probably never will be, but that is a tangent we shan't pursue here.) However, there is one principle of business, and more generally of the interchange of value between two or more human entities (meaning as small as one individual or as large as a *Fortune 500* corporation), that comes pretty close to being a "law." It does not have the mathematical clarity that Newton's laws do, and (unlike Newton's laws, within human distances and speeds) it is sometimes violated. That law, or principle, is shown in the accompanying box above.

Astute readers will immediately recognize this principle as simply a restatement, in different words, of the First Principle of Business Risk (§7.2), which notes that most humans are averse to risk. An observation in psychology translates into a great truth of finance and investment (and much more).

---

[10]Dynamicists have occasionally complained that Newton's law of gravitation is a bit difficult to work with because it is nonlinear. If business leaders could find anything as simple as $f_{12} = \mu m_1 m_2 / r^2_{12}$ to represent, say, the customer response to an advertising campaign, they would be shouting "Eureka" loud enough to damage the hearing of Archimedes (and he's been dead a long time).

[11]Except for some more recent modifications at hyper-teeny or breathtakingly vast distances, or at unimaginable speeds, or both.

Chapter 7: Some Risks Are Worth Taking

For some reason, Weak Form 2, even though logically equivalent to Weak Form 1, is more readily believed:

> **The Risk-Reward Principle (Weak Form 2)**
> Given two transaction opportunities, both with the same expected risk but with different expected rewards, most humans will take the one with the higher expected reward.

The weak forms of the risk-reward principle are interesting, but more realistically we must generalize to situations where the expectancies (expected rewards) and the expected risks are different.

Risk-reward Principle (Strong Form)

> **The Risk-Reward Principle (Strong Form)**
> Given two transaction opportunities, with the differing expected rewards and differing expected risks, most humans will tend to simultaneously reduce risk and increase reward to the largest extent possible.

When two propositions (e.g., investments) have different (expected) rewards and different (expected) risks, we get the "strong form" of the risk-reward principle, as shown in the accompanying box above.

This is more complicated, and Fig 7.7 should help. This plot shows a set of potential investments (indicated by dots) on a risk-reward plot. (The standard statistical symbols σ and μ are also used, but readers not helped by this connection to theory should just ignore these.) The strong form of the risk-reward principle implies that, of two investments actually available, indicated by A and B in Fig 7.7, A should be chosen because it is closest to the

---

[12] This boundary was called the "efficiency boundary" by Markowitz, in his March 1952 paper, "Portfolio Selection," in the *Journal of Finance*.

boundary[12] of possible risk-reward benefits, as determined by some larger market.

Figure 7.7: Risk-Reward Curve as Envelope to Possible Product Projects.

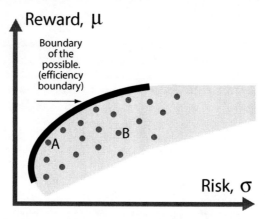

Although this diagram was first proposed as part of portfolio theory (and investing in the stock market), it applies equally well for a company that is continually reevaluating its portfolio of products. Note that the weak forms are special cases where the products A and B are either horizontally related (Weak Form 1) or vertically related (Weak Form 2).

Although we usually approximate the risk-reward curve (efficiency boundary) as a straight line with positive slope, and intersecting the vertical axis at the *risk-free rate of return*, we note that it actually can curve down especially if we use real returns (i.e., returns after inflation has been subtracted) so that the values before and after the return are in the same units ("before" dollars or "after" dollars). Thus, if Grandpa places his cash in his mattress because he thinks this is risk-free policy (which it isn't), the reward, after inflation, is actually negative.

Risk-reward Examples

As a first example of a risk-reward curve, consider the long-term risk-reward data for the ten asset classes shown in Fig 7.8. A best-fit straight line is also shown in the graph. The graph does have the general character of Fig 7.7.

Figure 7.8: Historical Risk-Reward Data for Ten Asset Classes.

Asset Class	Risk (Long-Term σ)	Return (Long-Term μ)
S&P 500	16.3%	11.7%
Small Stocks	25.5%	13.6%
T-Bills	3.2%	4.8%
US Gov't Bonds	9.7%	5.1%
Commercial Paper	3.6%	5.6%
US Farmland	7.4%	9.9%
Residential Housing	4.0%	7.2%
Venture capital	25.5%	15.9%
Foreign Equities	26.8%	13.2%
Inflation	3.8%	-4.4%

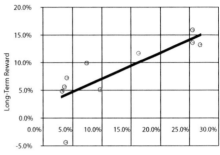

A similar diagram for mutual funds is given in Fig 7.9, showing that there is a fund (dozens, actually) for every risk preference.[13]

Figure 7.9: The Risk-Reward Relationship for Mutual Funds.

---

[13]Some readers may wonder at this point, "Isn't all this discussion off on a tangent? This isn't supposed to be a book about investing." No, it isn't. But consider that the angel investors in your startup could just as easily put their funds into one of the above mutual funds. Or in another startup with a better risk-reward profile. If your company is larger and public, your shareholders can sell their stock in your company and buy something else with lower risk or greater reward. And finally, if you are on the board of directors of a large, established company, and the company has significant retained earnings at the end of the year, and your board and senior management don't have any really good ideas for new business opportunities for your company, you should give at least some of these earnings back to your shareholders as dividends, so that they can get closer to the risk-reward equilibrium curve by putting their dollars somewhere else!

Note also the usual meaning of risk in this context—the long-term standard deviation, σ, of the data series. However, the author would much prefer[14] to use the word *volatility* in this situation—a word sometimes used. While the risk-reward relationship underpins all of investment, by funds, insurance companies, individuals, or other companies (the latter in capital projects and new product development, for example), the two applications of most interest in this book do not, unfortunately, have this kind of statistical basis. We have the stock performance data for General Electric or IBM over decades, but we (by definition) *have no data at all* on the past performance of a new product. We have experience with similar products, similar markets, similar companies, etc., but it's more a guessing game using the techniques explained earlier in this chapter to arrive at a risk estimate.

A second example of a situation of great interest to us is the startup company (Chapter 12). Again, we have no past data on the value of such a company; we do know that the risk is high, so the reward for investing in it should also be high. Thus the qualitative risk-reward relationship is still relevant, though not as precisely or quantifiably as with older, publicly-listed companies.

SWOTing Risks

There is one final exercise that is well worth doing, at least annually, in recognizing all the possible risks and rewards either for a particular product (or other company project) or for the company as a whole. This process is so obvious that it must have been done at least subconsciously for millennia, but it is currently known as SWOT Analysis for reasons obvious from Fig 7.10. This figure is not an x-y plot, but a "4-box" of possibilities.

---

[14]Short-term volatility, however, is indistinguishable from real risk.

Figure 7.10: The SWOT Approach to Assessing Potential Risks and Rewards.

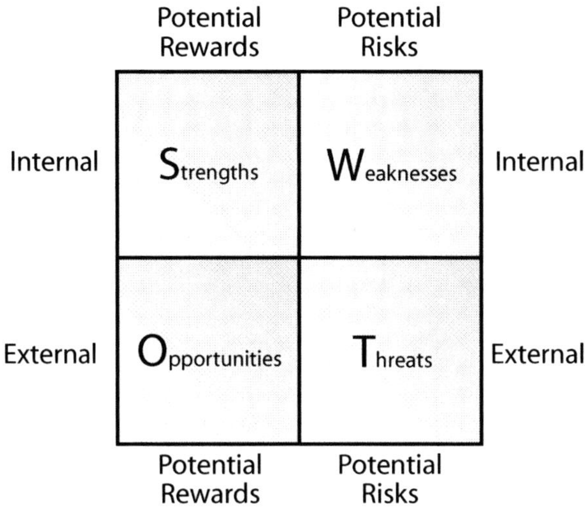

	Potential Rewards	Potential Risks	
Internal	S_{trengths}	W_{eaknesses}	Internal
External	O_{pportunities}	T_{hreats}	External
	Potential Rewards	Potential Risks	

Each of Strengths, Weaknesses, Opportunities and Threats starts off as a long list of items, drawn ideally from many knowledgeable people in the company. It may help further to subdivide the list into "technical," "market," "personnel," "management," "financial," etc., along the lines of the discussion earlier in this chapter. If people are honest and permitted to be honest, these lists can get quite long. Just the benefit of discussing all these pieces, and seeing them in black and white, is a salutary exercise. It should of course be kept confidential within the company. If one further wishes to quantify entries on these lists, even roughly, by associating numerical values with the most important items on these lists, one can then develop at least a semi-quantitative assessment of risk and reward.

CHAPTER 8

Accountancy

The Rules for Keeping Score

We shall not waste much time explaining why accounting is important in any organization. If one's organization is a for-profit corporation, then the bottom line is, well, the bottom line, and a great many calculations are required to arrive there. If one's organization is a not-for-profit organization, then the measurement of success is a more complex task, but this does not relieve it of the responsibility for keeping adequate financial records. So we shall take it as a given that accounting is important to any organization.[1]

8.1 ENGINEERS VIS-À-VIS ACCOUNTANTS

Why should engineers care about accounting? It seems self-evident that all (including engineers) who aspire to positions where P&L (meaning profit and loss) responsibility is involved—or higher—have to know what profit and loss is, how to boost the former and reduce the latter. Even lower down in the hierarchy, however, engineers doing basic engineering with little management responsibility should still have some ideas about accounting. Unless the

---
[1] Not quite as important as the accountants think it is, of course, but that's to be expected (from *vocophilia*, see §3.4).

organizations that employ them are set up as employee philanthropies—which for-profit corporations most assuredly are not—the organizations that pay these engineers their salaries and benefits, and that expend additional resources interacting with them and training them, are going to expect these engineers to create value greater than[2] all the costs of employing them. It behooves these engineers to have some idea of the financial principles that underlie their survival.

Engineers Don't Naturally Understand The Accounting Culture
But enough said about engineers who are not going to be managers and who don't wish to be managers. They have enormously challenging work to do, with vast benefits to themselves, their companies and society as a whole. They should know something about accounting, but they can't be expected to know much. This book is for engineers who are going to be managers or who are at least *thinking* about that career path, so the burden to learn something in the accounting area is no longer elective; it is mandatory.

Most engineers, by nature, begin by having a strained relationship with accounting. This is partly simple vocophilia (§3.4) but there are deeper reasons. First, both engineers and accountants spend a lot of their time fiddling around with numbers; frankly, neither can do their jobs for any length of time without some sort of calculation. The engineers, if they spend any time looking at accounting at all, catch on rather quickly to the fact that the mathematics used by accountants is rather trivial. In fact, it isn't really mathematics at all in any sense used by anyone out of elementary school.[3] It is, instead, simple arithmetic: addition, subtraction, multiplication and division of numbers (always to two decimal places).

---

[2] Why "greater than" and not just "equal to"? Because if the value the employee creates is just equal to the total cost expended on employing him or her, there's no point. Where's the profit? (We are talking about for-profit organizations.)
[3] We are distinguishing here between *accounting* and *finance*. The mathematics in accounting is quite trivial, but the mathematics in finance can be very challenging. In fact, it is not unusual for a mathematically talented engineering student to take an advanced degree in finance and thereby launch a fine career in that field.

Chapter 8: Accountancy

At this point, the engineers think that they have uncovered a scam. Accountants are (they think) quite overpaid because accountants' maths skills are primitive compared to theirs.

But this is to misunderstand the value added by accountants. They do not get paid for their arithmetic. They get paid for knowing, and applying for their clients, a quite labyrinthine set of rules.[4] It is a basic human fact that people's morals become malleable where money is concerned, and every one of the gazillion money scams that has ever been perpetrated has led to a counteracting rule. The aggregate of these rules is the stuff of accounting.

Then there are the tax collectors. People frequently talk about "government money" but the government has no money. In fact most governments are far into debt; they have to extract taxes from the productive individuals and organizations in their jurisdiction. The rules appertaining to this massive transfer of funds take up many volumes of small print—the tax code—and are constantly changing. Accountants make much of their fees and salaries by sorting all this out for clients.

A Subject With No Theory

There is a second major distinction between engineers and accountants which also leads to the strained relationship—their prevailing pedagogies are exactly the opposite of one another. Consult a typical engineering textbook, especially an applied science textbook. There is always a careful development of the general theory of the subject, meaning a quite general mathematical treatment. If space permits, there may be a few examples scattered about as well. *Courses* in applied science are taught the same way. A general theory of the subject is propounded in class, despite all the evidence that students (except the most mathematically gifted) learn best by going from the particular to the general. Students often criticize the lack of practical examples.

---

[4]In Canada and the USA, these rules are collectively called GAAP (Generally Accepted Accounting Principles)—one of the few acronyms from accounting that everyone should know.

Accounting is taught the converse way. Pick up any accounting book and one sees essentially a very long list of examples. The general theory is never presented—because *there is no general theory;* there is only GAAP, or the tax code, or some other bewilderingly long list of accounting rules. To most engineers, this kind of generalization is unexciting. Exciting or not, they must engage the subject to some degree because if they are going to be managers, they will be frequently speaking, directly or indirectly, with accountants. For an engineer to be buffaloed by what an accountant is saying is surely hard to explain (not to mention career limiting) if, as many engineers feel, engineers are the supposedly superior species!

Some Strengths And Limitations
Accounting, however, has its limitations, of which the engineer-manager must also be aware. Here, we mention two.

The first is forecasting, budget projections and the like. The first call on accountants is to record what *has* happened in a manner compliant with basic accounting principles. While anyone, including accountants, can attempt to predict the future by a straightforward extrapolation of the past—the math is, as usual, undemanding—simple extrapolation is almost like picking stocks by looking for strange meanderings in the stock price. Simple (usually linear) projections of the past are something an engineer can really relate to, but an engineer-manager knows that there is a lot more to accurate prediction that this.

For example, if bad weather has wiped out nearly all the coffee plants in Colombia last week, what good is a linear extrapolation of the past two good years in predicting the price of coffee? Looking at some simple line on a graph is merely an affectation of knowledge in a swamp of ignorance. Note carefully that it may not be the accountant's responsibility[5] to predict the future, so the fact that he refrains from doing so cannot be held against him.

---
[5]With special training, some species of accountants do become adept at making future projections. For example, *management* accountants, who are usually part of the firm, and who understand its functioning and business environment in intimate detail, can make intelligent predictions about the next few business years.

Chapter 8: Accountancy

A second situation in which accountants are only of partial assistance—and again it can't really be held against them—is what might be called a "cost-benefit" decision. The basic weakness is usually on the benefit side. The scenario usually unfolds in the following manner: management is thinking of doing something—hiring someone; firing someone; doubling the advertising budget; developing a new product; or installing a new computer system. There are thousands of generic examples. There will be a cost, and there will be a benefit. There may be some imponderables on the *cost* side, but these are usually dwarfed by the vagueness of the *benefit* calculation.

> **Cost-Benefit Decisions Take More Than "Accounting"**
> Many (rarely all) *costs* can be predicted with great accuracy. Financial *benefits* can almost never be predicted with accuracy.

Since the author has never seen anyone else remark on this phenomenon, it is highlighted in the accompanying box. Again, it's all about the ability to predict. The cost is usually a very-short-term prediction—perhaps as simple as looking up a price in a catalog—but the benefit is, in contrast, nebulous and typically accumulates over many years. So, who decides? Accountants can help, but it's a business management decision. As they say, that's why managers get the big bucks.

There are times when an accountant's advice is flouted at one's peril. A tax requirement. An offside bank covenant. A GAAP interpretation. One must have a very strong argument (though it can happen) to disagree with one's accounting colleagues on such matters. On other occasions—financial forecasts and cost-benefit riddles are two general examples just briefly mentioned—the accountant can give valorous assistance, but the final decision is a business decision.

8.2   AN EXTENDED EXAMPLE OF FINANCIAL ACCOUNTING

This section is written for engineers who have had no training at all in accounting. If the reader has had an accounting course (or picked up the equivalent elsewhere), please skip to the next section.

Presented below is an example, a story[6] in six short scenes, of financial accounting.[7] It builds up, in a rational manner, the purpose of such financial statements as the Income Statement and the Balance Sheet.

Dale's Lemonade Stand—Scene 1

Dale is an energetic, creative youngster who decides, in the summer of his twelfth year, that (a) it is hot outside, and (b) he doesn't have enough spending money. Since he is neither a whiner nor a beggar but a budding entrepreneur, he decides to set up a business selling lemonade. Accordingly, he buys some sugar and some lemons and sets up shop at a nearby residential intersection, selling Dale's Lemonade at $1 a glass.

At the end of the summer, Dale has to return to school and the demand for lemonade has fallen off dramatically. (Dale's Lemonade is a cyclical business.) At this point, Dale decides to assess the results of his summer operations and prepares the simple Income Statement shown in Fig 8.1. (Water was free.) He was pleased to see that he had made a profit.[8]

Figure 8.1: Income Statement for Dale's Lemonade (Try #1).

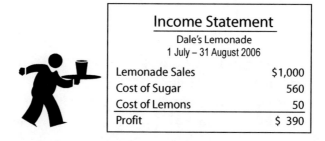

Income Statement	
Dale's Lemonade	
1 July – 31 August 2006	
Lemonade Sales	$1,000
Cost of Sugar	560
Cost of Lemons	50
Profit	$ 390

---

[6] Lest readers presume that considering a "lemonade stand" is beneath their dignity, consider the following item reported the week this chapter was being written, in the July 18, 2005, issue of *Maclean's* magazine: "[Nanaimo] orders Ana Cross, 10, to move rickety lemonade stand off municipal land. Little Ana gets valuable life lesson: If fate hands you lemons, make lemonade—but build your stand to code, get a license, and remember you can't fight city hall." Perhaps an allowance should have been made on the Income Statement below for expensive legal advice? This citation does have a point beyond human interest and humor. Anyone who tries to succeed in building a business should have a thick skin and a great deal of patience. Considering its overall benefits to our society, it is surprising to some that there are so many roadblocks.
[7] *Financial accounting* will be defined more carefully in §8.3.
[8] The terms "net profit," "net income," "bottom line," and "earnings" all refer to the same good thing.

Chapter 8: Accountancy

Dale's Lemonade Stand—Scene 2

Dale was excited about this success and exulted to his mother that he had worked hard and had made a profit. His mother was pleased and congratulated her son, but as it happened she had had some exposure to accounting, so she asked, "You have taken care to account for the consumables, but I can think of at least three items you used every day in your business that don't seem to have been accounted for. What about the trays? Or the glasses? Or the lemonade stand itself?"

Dale realized that one of the reasons he seemed to have made a profit was because he had left out some important expenditures. He had, indeed, purchased additional items to support his business. So he revised his accounting as shown in Fig 8.2.

Figure 8.2: Income Statement for Dale's Lemonade (Try #2).

Income Statement	
Dale's Lemonade	
1 July – 31 August 2006	
Lemonade Sales	$1,000
Cost of Sugar	560
Cost of Lemons	50
Stand	700
Glasses	400
Trays	300
Profit	-$1,010

Dale was not pleased with this accounting statement; it showed that he had not made a profit after all. He had, instead, lost[9] $1,010.

Dale's Lemonade Stand—Scene 3

Dale's father noticed his son's grim mood and asked about his lemonade results. "You are being a little hard on yourself, son," he said. "Granted, the sugar and lemons are all consumed, but you still have the lemonade stand, and the glasses and trays for next year. It's a bit

---

[9] All educated groups in the world—apparently with one exception—have heard of, and feel comfortable with, negative numbers. If profit is positive, loss must be negative. Only elementary-school arithmetic is required. However, for reasons that are inscrutable, accountants do not feel comfortable with negative numbers, possibly because they wish to distinguish between the state of negativity and the operation of subtraction. Thus, they don't write "–$1,010"; they write ($1,010) instead.

unfair to charge them all to this year's business activity. Will your lemonade business be a going concern next summer?"

"Yes, I plan on that," said Dale.

"Then carve out those longer-term items and list them separately. They're called *assets*. That will give you a better picture on the long-term business."

Dale complied, as shown in Fig 8.3.

Figure 8.3: Financial Statements for Dale's Lemonade (Try #3).

Income Statement	
Dale's Lemonade	
1 July – 31 August 2006	
Lemonade Sales	$1,000
Cost of Sugar	560
Cost of Lemons	50
Profit	$ 390

Assets	
Dale's Lemonade	
31 July 2006	
Stand	$ 700
Glasses	400
Trays	300
Total Assets	$ 1,400

Dale was quite pleased with this accounting slight of hand. He had apparently recovered his $390 profit, and now he also had a list of his business assets as well. He also noticed in passing that his Income Statement applied over a specified interval of time, while his Balance Sheet referred to the state at a particular instant in time.

Dale's Lemonade Stand—Scene 4

Dale's mother, ever sensitive to her son's mood, noticed the improvement and looked at his latest financial statements. "You're making progress," she said. "But I think your father was being too kind. You gave your business credit for three assets, yet there is no mention of how you acquired those assets. They didn't come from nowhere!"

Somewhat reluctantly, Dale recalled that his mother had, indeed, lent him $600 on July 1, and that his father had initially bought half the company for $700. He had said that he wasn't interested in making money off his son's labor, but that he was very willing to invest in his future. Dale was thus led to list, not only his assets, but his *liabilities* as well.

Figure 8.4: Opening Balance Sheet for Dale's Lemonade (Try #4).

Income Statement		Balance Sheet (Opening)			
Dale's Lemonade 1 July – 31 August 2006		Dale's Lemonade 1 July 2006			
Lemonade Sales	$1,000	Cash	$1,300	Debt	$ 600
Cost of Sugar	560			Equity	700
Cost of Lemons	50	Total Assets	$1,300	Debt + Equity	$1,300
Profit	$ 390				

The Income Statement was not changed by this adjustment, since it referred to the period following the opening Balance Sheet.

Dale's Lemonade Stand—Scene 5

Dale suddenly realized that his Balance Sheet would be changing constantly after July 1, as business operations carried on, and that he could, in particular, strike a new balance sheet for August 30. This, combined with his Income Statement over the same period, would give him a nice set of financial statements for his first year of operations, and a complete picture of the business (Fig 8.5).

Figure 8.5: Summer-End Financials for Dale's Lemonade (Try #5).

Income Statement		Balance Sheet			
Dale's Lemonade 1 July – 31 August 2006		Dale's Lemonade 31 August 2006			
Lemonade Sales	$1,000	Cash	$ 290	Debt	$ 600
Cost of Sugar	560	Stand	700	Equity	1,090
Cost of Lemons	50	Glasses	400		
Profit	$ 390	Trays	300		
		Total Assets	$1,690	Debt + Equity	$1,690

Dale noted that the equity was what was left over after the liabilities were deducted from the assets (what accountants might call the *net assets*, or the *assets net of liabilities*). He now felt he had completed his statements and was pleased that he was learning a lot about accounting. The cash account balance of $290 was what his bank statement indicated, but he also checked it against the fact that he had started with $1,300 cash, then had $1,000 in sales, $610 in expenses (as recorded on the Income Statement) and $1,400 in capital expenditures (for assets, listed on the Balance Sheet).

Dale's Lemonade Stand—Scene 6

That night at dinner, Dale gave his father the good news. "Dad, the company was worth $700 when you invested two months ago. Your shares were worth half[10] that ($350). Now, Dale's Lemonade is worth $1,090, of which your half is $545. That's a capital gain (on paper) of $195 for you. I guess it's not fair to assign all of that to just two months, since the company will be dormant until next July, but even over a year, that's still $390/$700—a 56% annual RoE (return on equity). And, as a manager, I achieved a RoA (return on assets) of $390/$1,300, or 30%. How 'bout them apples?"

"I think you mean 'How 'bout them lemons'," said his mother. But what happened to my $600 loan? You agreed to pay me 5% per annum. I haven't seen any interest yet, much less some paydown on the principal. And there's another thing you've left out. I noticed that a couple of your glasses were broken, your trays are a bit scratched, and your stand is somewhat weather-worn. I'm not blaming you for those things. They're a natural part of business operations. Assuming your assets will last no more than five years, I suggest you take a 20%/year (straight line) depreciation charge on your income statement for those assets. That would give a truer picture of your business."

"And another thing," said his father. "You started this business to make some spending money, but so far you've paid yourself nothing. You won't find too many employees like that. And if, after all of this, the company still made some money, you will have to set aside 25% of your profit for taxes. Let's see how all that turns out."

---

[10]Note that, in effect, Dale's father had, at the outset, given his son $350 in value. While he had put $700 into the company, he owned only half of that (half the equity of $700). However, this is not the straight paternal gift one might first assume. Dale's ownership was based, not on his contribution of capital, but on the fact that the company was his idea, and he was doing all the setup work to get it started—without pay. In other words, he was a *founder,* and founders are often acknowledged by getting just what Dale got: *sweat equity.* More on this in Chapter 12.

Chapter 8: Accountancy

Dale went back to his desk once more. He was a bit afraid of what he would find after paying his mother the interest he owed her, allowing for the depreciation that had occurred, and paying himself and his secret partner (the government).

Dale began to realize (Fig 8.6) that "making money" was not all that easy. He had only paid himself a dollar a day (one glass of lemonade); he had not repaid any of the principal on the loan to his mother; and he had not refurbished his assets (glasses, stand, etc.)—and yet he still almost lost money.

Figure 8.6: Summer-End Financials for Dale's Lemonade.

Income Statement			Balance Sheet			
Dale's Lemonade 1 July – 31 August 2006			Dale's Lemonade 31 August 2006			
Lemonade Sales	$1,000		Cash	$ 195	Debt	$ 600
Cost of Sugar	560		Stand	560	Equity	715
Cost of Lemons	50		Glasses	320		
Gross Profit	$ 390		Trays	240		
Payroll	60		Total Assets	$1,315	Debt + Equity	$1,315
Depreciation	280					
Interest on Debt	30					
Earnings before taxes	$ 20					
Taxes	5					
Net Income	$ 15					

As a check on shareholders' equity, Dale set up the following statement (Fig 8.7):

Figure 8.7: Statement of Shareholders' Equity for Dale's Lemonade.

Shareholders' Equity	
Dale's Lemonade 31 August 2006	
Shareholders' Equity [beginning of period]	$ 700
Retained Earnings [during period]	15
Shareholders' Equity [end of period]	$ 715

Since no dividends were paid out, the retained earnings were the entire net income during the period. His RoA had been $15/$1,300 = 1.2%, and his RoE had been $15/$700 = 2.1%. Dale decided to spend some time during the colder seasons to formulate a detailed business plan for next year.

It is hoped that much of the rationale behind Income Statements and Balance Sheets will become apparent from the above extended example, along with some of the more important items of accounting terminology.

8.3 FINANCIAL AND MANAGERIAL ACCOUNTING

With the introduction provided by Dale's Lemonade in the last section (or with an equivalent accounting background), we can now look at financial accounting in at least *some* greater generality. We can't be completely general, because, as said earlier, there is no general theory. Large public corporations will have financial statements that are much more complex than those we'll be looking at. Nevertheless, the understanding provided below should give considerable insight, so that when an engineering manager[11] looks at a balance sheet he or she won't feel like, say, an accountant looking at an algebra book.

Financial Reporting Is Done For Outsiders

It is time to remark on the rather strange term, "financial reporting." A novice to the craft of accounting must surely be asking, "Since accounting seems to always involve dollars, isn't all of accounting financial? And aren't *all* accounting statements reporting something?" The correct answers are, respectively, "Yes" and "Yes." Yet the term *financial reporting*, as used by accountants, refers only to a relatively small subset of all the possible dollar-based accounting

---

[11]Meaning, as should be obvious, an engineering manager who has not studied accounting. This is written as an overview for those who have not yet done so.

reports. Why is this? The only answer that seems plausible is that, as with all disciplines and specialties, especially old[12] ones like accounting, the words used by its practitioners have shades of meaning different from those in the everyday vernacular.[13]

The meaning of financial reporting may be clarified with the aid of the accompanying box and Fig 8.8. *Managerial* accounting is one of many activities always going on within any corporation. All of *financial* accounting deals with outputs[14] from internal (i.e., managerial) accounting, subject to all the rules mentioned previously. There is no secret, additional ingredient.

"Financial" Accounting
Statements are for *external* consumption—banks, investors, shareholders, securities regulators, etc.

"Managerial" Accounting
Activities are for *internal* consumption—planning, decisionmaking, directing and controlling, etc.

---

[12] Accounting of some sort has been around as long as humans have engaged in financial transactions. For example, the word *auditor*, originally meaning "someone who listens attentively," was applied to accounting because, in ancient times, when few were literate, the financial state of an organization was read out loud to members of the public, who listened attentively. Nowadays, auditing is a specialization within accounting.

[13] Engineers and scientists cannot criticize accounting, or any other profession, for quirky use of words. For example, scientists refer to the "Theory of Gravitation," by which they mean a mathematical formulation that is so precise it has enabled scientists and engineers at NASA's Jet Propulsion Laboratory, the very week these words are being written, to use a man-made spacecraft called Deep Impact to send a small interceptor to crash successfully into the comet *Temple 1* (relative speed = 23,000 mph). Described as "hitting a bullet with a bullet with a third bullet taking pictures," this accomplishment is just the latest triumph for the Theory of Gravitation. In contrast, popular argot uses the word "theory" as in "That's just another theory, but here's my theory." Nothing special; just an opinion. So we end up with people who want creationism taught in science class as an alternative to the Theory of Evolution. Why not? They're all just theories. To a scientist, of course, and to an engineer with solid science training, the Theory of Evolution is comparable to the Theory of Gravitation. It is the same specialized use of the word Theory. (On the other hand, if creationists were suddenly exercised to create their own theory of gravitation, and with the same level of skill as their theory of creation, they wouldn't be able to predict whether a bible would fall up or down when dropped from a pulpit.) Compared to this enormous divide between the specialized and the popular meaning for the word theory, the accountant's distinctive use of financial statements or financial reporting is negligible.

[14] *Outputs* as defined in *system theory*, for engineers who have studied the subject, which most recent graduates have.

Figure 8.8: The Inside-Outside Distinction.

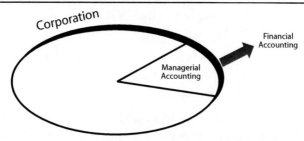

We shall arrive at managerial accounting presently, in §8.7. In the meantime, we note that there are few rules imposed on managerial accounting (intended for internal consumption) but that so-called financial reporting is subject to GAAP, the tax code, and a larger code of ethics (as mentioned earlier).

Managerial Accounting Is Done For Insiders

One has only to glance at the relative brevity of a company's *financial statements* to realize that a mountain of financial data generally underlies their preparation, and that the larger the company, the larger the mountain. Unless one's company is as small as Dale's Lemonade, one cannot use a single accounting for both inside and outside uses. For companies of even modest size the official financial statements for public consumption represent just the tiniest tip of a very large iceberg. For the largest companies, even the iceberg analogy breaks down; there are hundreds of offices filled with managerial accounting information. We shall address managerial accounting more completely in §8.7.

8.4 THE INCOME STATEMENT

The example of how the Income Statement for Dale's Lemonade came to be invented will, one hopes, provide some motivation for its structure and purpose. Here, we generalize the Dale's Lemonade model to some extent, while recognizing that, in full flower, Income Statements must be prepared by professional accountants, as appropriate to corporations involved. A generic model is shown in Fig 8.9.

Figure 8.9: A Generic Income Statement.

Income Statement	
ABC Company	
1 January - 31 December 2006	
(in $000,000)	
Revenue	$1,200
Cost of Goods Sold	350
Gross Profit	$ 850
Marketing	200
General & Admin	100
Research & Development	50
EBITDA	$ 500
Extraordinary Item	100
Depreciation	80
Amortization	50
EBIT	$ 270
Interest on Debt	50
EBT (Operating Profit)	$ 220
Taxes	30
Earnings (Net Income)	$ 190

The company year happens to be the calendar year, but this is not necessary. The comments below pertain to each line in the above financial statement.

Revenue: Roughly the same as sales but usually with a time lag. A salesperson will expect their commission as soon as there is a document from the customer confirming the sale, but if the item has yet to be built (much less shipped, invoiced, and later, payment received) revenue will not be recognized (i.e., logged in as such in the revenue account) until somewhat later. The company should have a GAAP-compliant revenue recognition policy stated in the Notes to the financial statements, and should try to stick to the same policy year after year.

Cost of Goods Sold (COGS): The COGS are costs (expenses) that can be tied directly to individual items produced, including raw materials, manufacturing labor and any other factor that can be associated with producing finished goods. Heat and light, for example, are not part of COGS because if 10% fewer items had been produced, the heat and light would be unaffected, while the COGS should go down by 10%. COGS is a good example of a *variable cost*, whereas heat and light tend to be *fixed costs*.[15] Some use the term COS (Cost of Sales) in place of COGS. This might be done, for example, if the company sells primarily services, not goods. However, the COS terminology can sometimes be confusing to a learner since sales is part of marketing, yet marketing expenses occur later on the Income Statement. In the example of Dale's Lemonade (§8.2), the COGS was based on the cost of the sugar and lemons used.

Gross Profit: Defined as what is left over after deducting COGS from revenue. The word "gross" is connected to the fact that this word was originally the French word *gros* meaning large. The Gross Profit is the large profit because there are many expenses still to be deducted.

Marketing: The general meaning of these expenses is self-explanatory. A glance at Chapter 5 should provide some insight here. Included are both marketing and sales expenses. (Sales is part of marketing.)

General and Administrative Expenses (G&A): This item covers a wide territory and is in fact something of a catch-all, including rent, management salaries, heat, light, telephone, office supplies and staff, and on and on—and of course we can't forget legal and accounting fees.

Research and Development (R&D): Many companies would not have this as a line item in a top-level (i.e., undetailed) Income Statement such as the one shown above in Fig 8.9. However, we have done so here, as high-tech companies sometimes do, because

---

[15]Nothing is truly fixed with mathematical precision. For example, if production is doubled, suppliers may give a discount on the raw materials. If there is a new wage settlement in the middle of the year, unit labor costs may go up a bit. If business is brisk and the manufacturing floor goes from one shift to two, the heat and light may go up somewhat also, and so on.

Chapter 8: Accountancy

our main readership is engineers—and some of these readers may be doing R&D for companies.[16]

Earnings Before Interest, Taxes, Depreciation and Amortization (EBITDA): See subsection to follow.

Extraordinary Item: This expense is separately classified because it would arguably not be an expense in normal business operations. For example, losing a large lawsuit or suffering high damage from a hurricane would likely qualify as an extraordinary item. (Some accountants use the term *Unusual Item*.) In any case, it must be explained in the Notes to the Financial Statements. The whole idea is to explain why profit is not higher. Of course, if a company has some Extraordinary Item three years out of five, these items have become rather ordinary; shareholders and valuators can make their own inferences.

Depreciation: We have already seen an example of depreciation with Dale's Lemonade. There are several related ways to look at the foundational meaning of *depreciation*. In one of these the accounting treatment mirrors the physical fact of deterioration, and since the value of an object generally depreciates (lessens) along with its physical[17] deterioration, its value on the Balance Sheet should be reduced, year by year.

The second way to view depreciation is as an allocation of the purchase cost of an asset over its life. Moreover, this loss in value has to go somewhere, and the only place it can go is as a charge on the Income Statement.

---

[16]This R&D item leads to remarking on an interesting difference between Canadian GAAP and American GAAP. In the latter, R&D must be "expensed," i.e., shown on the Income Statement as an expense (Fig 8.9). In Canadian GAAP, one has the option of *capitalizing* R&D expenditures, i.e., placing them on the Balance Sheet as an (intangible) asset, to be depreciated over a period of years, based on the argument that the benefit of the R&D will occur over several years and so the corresponding *expenditure* should also be taken onto the Income Statement over the same several years, via depreciation of the R&D asset. The latter does make sense if one is willing to believe that all, or a high fraction of, the company's R&D *does* reliably lead to revenue in future years. Apparently, U.S. accountants have their doubts about a strong, reliable connection between R&D and future revenue—and this is *also* a reasonable view. (Notice the word *expenditure*, which, in accounting, is more general than *expense*; the latter is used only for Income Statement items.)

[17]This would not apply to a newly found da Vinci painting or to an original score of Mozart's Symphony #42 hiding in an old Vienna museum.

The third meaning of depreciation emphasizes the concept of *matching periods*. If a major purchase, such as an expensive BigWidget, is made, with a view to improving the business over the next N years, it seems only fair to attribute that cost equally over the business operations of the next N years (i.e., over the next N annual Income Statements), and decreasing its value on the Balance Sheet accordingly as the years go by. This period of depreciation may not be identical to the period of physical deterioration. This meaning is more general than the one emphasizing physical deterioration because the asset does not actually have to be a physical thing for the reasoning to apply.[18] Thus, in Canadian GAAP one can depreciate earlier R&D expenses, which are not tangible.

A fourth use of depreciation is as a device to manage earnings. Clearly, other things being equal, the larger the depreciation in a particular year, the less the earnings that year. Less income means less tax, although the tax department has its own rules about how fast a company can depreciate its assets. On the other hand, a smaller rate of depreciation will swell the earnings, attracting the favorable attention of one's banker and one's shareholders.

Amortization: Very similar in spirit to depreciation—in that it spreads a major capital expenditure over several years—except that the asset in question is financed by *debt*, or is an *intangible* long-lived asset.[19]

Earnings before Interest and Taxes (EBIT): This is self explanatory from Fig 8.9.

Interest on Debt: Managers are attracted to debt because they believe that they can generate a RoA higher than the interest on debt. If they are right, this approach is fine. If the company runs into trouble, however, debt can become a heavy load. Not surprisingly, lenders (banks and bond-holders) limit the allowable debt/

---

[18]Compare with the discussion of the word *good* in §5.2. A depreciable asset does not have to be a good.
[19]The meaning of amortization within accounting is not perfectly defined. No engineer or scientist would feel comfortable with such vagueness within his or her profession.

Chapter 8: Accountancy

equity ratio through *restrictive covenants* before they are willing to advance funds.

Earnings before Tax (EBT) (a.k.a. Operating Profit): Unlike most of the terms we are discussing, the term *operating profit* is hard to pin down with precision; it has several slightly different, though related, meanings. These tend to vary by industry. For example, the financial sector has its own definition of operating profit. For the purposes of this chapter—making engineers feel more at home with accounting, but not trying to turn them into accountants—the definition implied in Fig 8.9 will do just fine.

Taxes: This is the one we all know about, in principle, although the details for corporations keep a major army of accountants in business.

(Net) Income, or Earnings: The infamous "bottom line," whose general meaning[20] is clear from Fig 8.9.

EBITDA—Just A Scam?

Lately it has been popular in accounting circles to talk a great deal about EBITDA, and much of this talk is rather silly and seems to be a third-rate scam.[21]

> Some accountants pronounce EBITDA as EE-BIT-DAW; others pronounce it as EE-BIT-DAW. The latter sounds better on weekends.
>
> EBITDA stands for Earnings Before Interest, Taxes, Depreciation and Amortization, that is, before a lot of items have been deducted. Unsurprisingly, this gives managers a warm feeling about how big this sort of earnings is, which is why they play along. The accountants, summing up their most solemn look, say things like, "Of course, de-

---

[20] Note that the word "income" for a corporation has a meaning quite different from personal income. Personal income for individuals corresponds more closely to "revenue" for corporations, and income for corporations is similar to savings for individuals. The top line and the bottom line can easily be confused the first time around.

[21] By third-rate scam, the author means a scam in which the scamee willingly participates all along, even knowing that it is a scam. A *second*-rate scam is one where the scamee *eventually* knows he's been scammed, but it's too late. A *first*-rate scam is one where the scamee *never* knows he's been scammed! (For an excellent example of a flawlessly executed first-rate scam, watch Robert Redford and Paul Newman in the movie *The Sting*.)

preciation and amortization are not cash expenses," and the managers nod their affirmation as though something profound has just been said. As though being not cash somehow means not real. But they are real, and in a strong sense they are also cash.

Time for an analogy. On December 1, a young couple buys a new home with a 20-year mortgage. A week later, they fill it with fine furniture on the layaway plan, whose terms are that they are to make equal monthly payments for five years. On January 1 of the following year—that is, one month later—Hubby says to Devoted Wife, "You know, dear, we're doing very well financially. It looks like we'll have a pretty good cash position at the end of this month."

"How can that be?" asks Devoted Wife, "What with the mortgage and the furniture payments?"

"Oh, I didn't include those," says Hubby.

EBITDA is almost that bad. Someone may say, "Well, in the case of depreciation, the cash has been fully paid for the asset." Actually, Hubby and Devoted Wife also fully paid in cash for their asset (their home), although they had to borrow a truckload of money (their mortgage) to do it. Most of their monthly mortgage payment is interest. Can they just neglect this interest when they assess their financial status? EBITDA does; it's nicely before interest. Perhaps Hubby and Devoted Wife can forget about their personal income tax when they examine how well they're doing? Not likely. But EBITDA does; it's also nicely before corporate income tax. And if they do not make their payments on the furniture, the repo truck will be backing up to their door. But EBITDA would ignore those payments.

If a company decides in Year 1 to purchase a big asset and to spread that expense (i.e., depreciate the asset) over five years on the grounds that the asset will boost business operations for five years, that's fine. But if, in Year 2, someone tries to claim that depreciation can be conveniently ignored in assessing the financial performance of the company because it's non-cash, that's not fine. The ink may be dry on the check, written over a year ago, but it's still cash—old cash perhaps, but cash nonetheless.

Furthermore, unless the company decides to let an asset depreciate to zero and then never replace it, there will be a current expenditure to

Chapter 8: Accountancy

upgrade or improve that asset, to compensate for its (real, physical) depreciation, and this expenditure (conveniently) doesn't show up on the Income Statement because it will also be capitalized as an asset and then depreciated over time. And those depreciations will then be ignored, and so on.

If some reader is still unconvinced that EBITDA is pretty much a scam, try this thought experiment. Instead of *annual* financial statements, let's have complete, *monthly*, audited financial statements. Make the fundamental accounting period one month instead of one year. That way, one can start talking about non-cash depreciation and amortization as early as *one month* after the commitments have been made. Why not? It's a simple two-step. First call them non-cash; then never mention them again in discussions of financial vitality.

As a last gasp at retrieving some semblance of respectability for EBITDA, a True-EBITDA-Believer may say, "EBITDA can be used to analyze the relative profitability between companies and industries. Because it eliminates the effects of financing and accounting decisions, EBITDA can provide a relatively good apples-to-apples comparison. For example, EBITDA as a percent of sales (the higher the ratio, the higher the profitability) can be used to find companies that are the most efficient operators in an industry." Well, these comments are helpful; they suggest that there is a specialized use of EBITDA that can reveal some truth.

The True-EBITDA-Believer presses on. "EBITDA can also be a good measure of cash flow." There are two responses to this claim. First, it is weak because, for many companies, EBITDA is *not* a good measure of cash flow; and, second, there is a financial statement called the Statement of Cash Flow (which we shall examine briefly below) which does, and is intended to do, precisely this function. Why introduce spurious approximations when they are entirely unnecessary?

The bottom line: With the exception of the very technical use mentioned two paragraphs ago—which is an unlikely activity for any engineer-manager—EBITDA is just a sexy-sounding acronym whose bloated magnitude, however unrealistic, is intended to make hard-working CFOs feel good.

219

8.5 THE BALANCE SHEET

The example of how the Balance Sheet for Dale's Lemonade came to be will, one hopes, provide some motivation for its structure and purpose.[22] Here, we offer a somewhat more general model (Fig 8.10), while recognizing, once again, that any actual Balance Sheet must be prepared by a professional accountant. The comments below apply to the major Balance Sheet categories:

Figure 8.10: A Generic Balance Sheet (no relationship to Fig 8.9).

<table>
<tr><td colspan="4" align="center">Balance Sheet
ABC Company
1 January - 31 December 2006
(in $000,000)</td></tr>
<tr><td>Cash</td><td>$ 50</td><td>Accounts Payable</td><td>$ 300</td></tr>
<tr><td>Accounts Receivable</td><td>400</td><td>Operating Line of Credit</td><td>300</td></tr>
<tr><td>Inventories</td><td>450</td><td>Short-Term Liabilities</td><td>$ 600</td></tr>
<tr><td>Short-Term Assets</td><td>$ 900</td><td></td><td></td></tr>
<tr><td></td><td></td><td>Long-Term Debt</td><td>$ 250</td></tr>
<tr><td>Long-Term Assets (Tangible)</td><td>$ 500</td><td>Total Long-Term Liabilities</td><td>$ 250</td></tr>
<tr><td>Long-Term Assets (Intangible)</td><td>350</td><td></td><td></td></tr>
<tr><td>Total Long-Term Assets</td><td>$ 850</td><td>Total Liabilities</td><td>$ 850</td></tr>
<tr><td></td><td></td><td>Share Capital</td><td>$ 250</td></tr>
<tr><td></td><td></td><td>Retained Earnings</td><td>650</td></tr>
<tr><td></td><td></td><td>Shareholders' Equity</td><td>$ 900</td></tr>
<tr><td>Total Assets</td><td>$1,750</td><td>Total Liabilities +
Shareholders' Equity</td><td>$1,750</td></tr>
</table>

Cash: In normal conversation, *cash* refers to what is in one's billfold, wallet or purse. To a business, cash means, primarily, the balance in its current account at the bank.[23] Modern Balance Sheets also

---
[22] To the timeless question "Do you want the good news first or the bad news first?" accountancy has decided that the good news should come first. Assets first; then, liabilities.
[23] Yes, for the fussy at heart, cash also includes the contents of that little coffee-money box operated by the receptionist, assuming that the legal tender and coins therein contained are contributed by the company and not by the employees.

Chapter 8: Accountancy

include in the general "cash" category so-called *near-cash*, meaning assets (such as marketable securities, government bonds, etc.) that can be converted to cash virtually immediately.[24]

Liquid: An asset is said to be *liquid* if it can be converted to cash quite quickly without penalty. Liquid *debts* can similarily be paid off quite quickly without penalty. Assets and liabilities come in all degrees of liquidity, the least liquid being called *frozen*. For, example, if one cannot get out of (pay off) one's home mortgage without paying three years' interest, that's quite frozen.

A *financial position* is said to be *liquid* if there is enough cash to purchase essentials right away. This is closely connected with the idea of *working capital* (see below). An old couple whose entire net worth is tied up in a paid-off home, but which has no cash for everyday necessities, has too little liquidity. A stock is said to be *liquid* if there are always many buyers and many sellers.

Accounts Receivable: A very general category meaning all the money owed to the company in the near future for goods and services rendered and for which it has billed. In short, it is the money owed by customers.[25]

Inventories: In the classical manufacturing scenario, there are three inventories, corresponding to the three longest and most unavoidable pauses in the manufacturing[26] process, as shown in the accompanying box.

---

[24] Engineering and accountancy both have well-developed vocabularies. The largest challenge comes, not when these vocabularies are different, but when they overlap. In the present instance we have the word "immediately." What does an engineer mean by this word? A geological engineer may mean "in the next century" while an electronics engineer may mean a nanosecond, two ideas that vary by 18 orders of magnitude. *Immediately* as used by accountants for the test of what is and what isn't cash sometimes permits conversion periods of up to 90 days for a marketable security to still be given the near-cash designation and thus listed as cash on the Balance Sheet.

[25] Note that the plural of "account receivable" is "accounts receivable," with the plural on the noun, where it should be. (Think of Attorneys General.) Sometimes, to shorten the category, the adjective "receivable" is used as a noun and one speaks of "receivables" in place of accounts receivable. But under no circumstances, except the desire to appear pitiful at both accounting and the English language, should one use "account receivables" as the plural.

[26] The manufacturing choice of example is made, not because it is all-important (although manufacturing is of long-term importance to any national economy), but because it is so rich in its list of examples of the inventory concept.

> **The Three Classical Manufacturing Inventories**
> [in chronological order]
>
> 1. Raw-materials inventory
> 2. Work-in-process inventory
> 3. Finished-goods inventory

These ideas must, as always, be applied with some intelligence. For example, if a company is involved primarily with system assembly and integration, one must substitute "subsystem" for "raw materials" and "integrated" for "manufactured." These will be discussed more fully in §8.7.

Short-Term Assets: The three examples of assets thus far (cash, accounts receivable, inventories) are all short-term assets, meaning[27] that they are expected to be converted into cash within a year. These categories change rapidly, in principle, possibly thousands of times per day for a large company (although the accounting is not done that often, of course).

Long-Term Assets: The distinction is with respect to short-term assets. Long-term assets[28] are held for a long while, sometimes for decades, and are relatively difficult to buy and sell. Figure 8.11 shows a plot vs. time of both a short-term asset and a long-term asset in the first four hours of a typical business day. The short-term asset (e.g., cash) is changing constantly as business operations and transactions progress, but the long-term asset value is changed very infrequently (e.g., once-per-year or once-per-quarter depreciation, or some occasional upgrade to the asset).

---

[27]Until recently, short-term assets were called *current* assets, an unhelpful term owing to the several meanings of "current." It could mean "now" or "up to date," both of which apply to *all* assets. Actually, it meant "sometime this year." *Short-term* covers the meaning nicely, especially for those with engineering sensibilities, although it will take a while for *current assets* to die completely.

[28]Until recently—and still likely far from extinct—the term "fixed assets" was used for long-term assets. Fixed in the ground? Fixed on the wall? They certainly weren't fixed in value, because the same old textbooks that introduced them as "fixed" always went on in the next paragraph to describe how they were depreciated—in other words, how their values were diminishing! Engineers tend to be upset by such contradictory terminology, but fortunately the modifier *long-term* is catching on rapidly.

Chapter 8: Accountancy

Figure 8.11: Fluctuations in Value of Short-Term Assets.

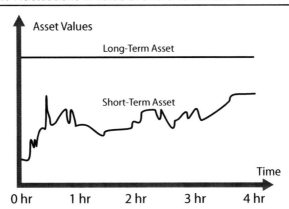

Tangible and Intangible Assets: Tangible means "touchable" and hence the distinction made is between physical items (assets) and more abstract ones. Land, buildings and machinery are quite *tangible* assets, although there is always the question of who else could use the machinery. Physical inventories are tangible although the bank will likely discount[29] their value somewhat. Banks are even more leery[30] of *intangible* assets, meaning that they are even more loath to lend money based on the going-forward business worth of these assets.

Accounts Payable: This is the mirror image of accounts receivable (and with the same warnings about syntax). In short, it is the money owed to suppliers.

---

[29]The *finished goods inventory* seems tangible enough, but if the company is winding up, who will actually sell these items? If, on the other hand, the company seems stable and is expected to be a going concern for the foreseeable future, the bank's discounting of the value of finished goods will be less severe. Similarly, the *work-in-process inventory* is valuable if the company continues as a going concern, but if it is going out of business, who will convert this work-in-process to finished goods? Likely answer: Nobody. What is the consequent worth of this inventory in this situation? Scrap at best. As for the *materials* inventory, it might be returnable to its suppliers at a discount, but someone has to actually look after doing that, leading to another cost chargeable against this asset from the bank's point of view.

[30]The tech boom, although having many regrettable impacts, did cause lending institutions to re-examine their attitude towards heavily *knowledge-based companies*, meaning companies whose assets were largely intangible. One example has been given earlier: Research and development—mentioned here because of its natural interest to engineers—is a permissible though intangible asset in Canada. But how much of the costly R&D effort leads to an acceptable RoA? How much leads to anything at all? The bank has a point. Patents or trademarks and, more generally, intellectual property (see Chapter 10) are further examples of intangible assets. As a very positive example, the value of the Coca-Cola brand name has been estimated by *Forbes* magazine at US$55B.

Operating Line of Credit (LoC): To ensure adequate working capital (see next definition) a line-of-credit loan can be provided by one's lending institution (likely a bank). The outstanding balance, in theory, bounces up and down hundreds of times a day as the multitude of everyday financial transactions take place (Fig 8.12). The bank will place a ceiling on this LoC, meaning that if a firm needs more working capital than the LoC ceiling permits, it must petition the bank to raise the ceiling. The bank also takes a dim view if the company never, even once per year, completely (and however temporarily) pays off its LoC entirely. The minimally unpaid amount, by definition, does not qualify as a LoC since it is apparently not short-term (meaning that it has, in accounting, a "best before date" of less than one year).

Figure 8.12: Fluctuations in the Line of Credit (and its Impermissible Long-Term-Debt Component).

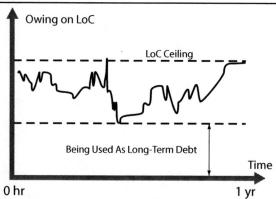

Working Capital: This could be called *short-term capital* because it is defined simply as

$$WC = STA - STL$$

That is, *working capital* is defined as short-term assets (includes cash) minus short-term liabilities. Each company wants to pay its suppliers as slowly as possible (thus using them as little banks), while insisting on payment from its suppliers as quickly as possible (thus preventing all of them from using the company as a little bank). This leads to a never-ending dance of negotiation up and down the value chain (supply chain).

Long-Term Debt: Self-explanatory. The interest on LTD is an expense on the Income Statement (as are all interests on all debts). If a certain amount of principal is due every year, this is called the *current (i.e., short-term) portion* and is listed as a short-term liability.

Share Capital, and Retained Earnings: Share capital are the funds that have been invested in the company by investors at various times in the past, and retained earnings represents the algebraic sum of all earnings (±) since incorporation. This distinction is carefully made to prevent Ponzi[31] schemes.

Shareholder's Equity: This is self-evident from Fig 8.10. The left side of the Balance Sheet shows all the assets owned, and the right side shows who owns them. The two sides must balance. The right side is a list of claims, or equities, against the assets of the company. However, in practice, the word *equity* is normally used only for the claims of *shareholders*.

Book Value: Closely related to shareholders' equity (shown on the Balance Sheet) is the *book value*, where reference is being made to the company's accounting books. It is the shareholders' equity minus any intangible[32] assets. For this reason, book value is also often called the *net tangible assets*. The actual value of the company (in a market sense) may be quite different again and will be studied later, in §10.1.

---

[31]Charles Ponzi (1882–1949) perfected a basic scam still often used today. He would collect investments from a group of credulists and then give them a fine return on their investments. These "returns" were, of course, merely the investments from an ever newer and larger group of investors. (Need it be stated? He also paid himself well for his own trouble.) If readers are ever asked to contribute to any financial arrangement of which they are suspicious, demanding a set of audited financial statements usually makes the perpetrator move on to lower-maintenance victims.

[32]Perhaps it should be explicitly stated that only assets can be intangible. A slip of paper from the bank stating that your company owes them a million dollars may not seem all that tangible, using the feel test, but material from the bank and the tax department is very tangible indeed.

Risk And The Balance Sheet

The configuration of the Balance Sheet presents an excellent opportunity to relate this material to the discussion of risk that occupied Chapter 7. We have already seen that short-term (the most active) accounts are placed above the long-term (and least active) ones. Now we observe, particularly with respect to the right side of the Balance Sheet, that the equities (owned by suppliers, lenders, and shareholders) are generally also listed in increased order of risk.

In Fig 8.13 we use A for Assets, D for Debts (i.e., liabilities) and E for shareholders' Equity. On the liability side, at one extreme, we have the obligation to pay employees and taxes—which is so mandatory that directors on the Board of Directors are personally responsible for ensuring that these are paid (see Chapter 13). Bank debt is secured, i.e., lent with assets as collateral. This is also called *senior* debt to distinguish it from *junior* debt which is less secured (or unsecured). Senior debt means "paid off first" in the event of company liquidation or bankruptcy. Thus senior debt-holders carry less risk and accordingly have less reward (interest rate) than junior debt-holders, who operate higher up on the risk-reward curve (Fig 7.7).

Figure 8.13: Risk Pattern to Equities.

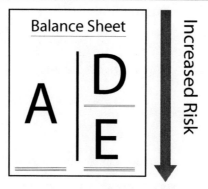

Taking the greatest risk of all are shareholders, but their equities will also reap the largest reward if things go well. If the company folds, however, the shareholders may be left with nothing, thus losing their entire investment.

8.6 CASH FLOW STATEMENT; AND NOTES

The Income Statement (§8.4) and the Balance Sheet (§8.5) are two legs of a three-legged stool that comprises the annual financial statements of a corporation. The present section explains briefly the third leg—the Cash Flow[33] Statement (Fig 8.14).

Figure 8.14: A Generic Cash Flow Statement [unrelated to Fig 8.9 or Fig 8.10].

```
               Cash Flow Statement
                    ABC Company
              1 January - 31 December 2006
                    (in $000,000)
Cash (as of 1 Jan 06)                    $  35

              Cash from Operations
Earnings (Net Income)                    $1,200
Cash Inflows & Outflows                     -50
Net Cash from Operations                 $1,150

              Cash from Investing
Cash Inflows & Outflows                  -$ 900
Net Cash from Investing                  -$ 900

              Cash from Financing
Cash Inflows and Outflows                $   10
Net Cash from Financing                  $   10

Cash (All Sources, 2006)                 $  260

Cash (as of 31 Dec 06)                   $  295
```

---

[33] Until recently (and perhaps still by some) the Cash Flow Statement was called the Statement of Changes in Financial Position—an unhelpful piece of jargon if ever there was one. Aren't all financial statements about financial position (whatever that is)? Do they not all record changes that have occurred? Exactly. The rationalization of some of the ancient but suboptimal terminology in accountancy has probably been due to the influence of a much younger subject—finance—which is very mathematical (and therefore very logical).

One begins with the cash left over last year (including dividends to shareholders); calculates the cash sources and sinks (there can be only three types: *operations, investing,* and *financing*) for all of this year; and then arrives at the cash left over at the end of the year. (Some of this leftover cash may be declared as dividends by the Board of Directors; see §13.3). A complete discussion of the Cash Flow Statement is beyond the intended scope of this book. (This explains why the totals in Fig 8.14 seem all to have just one item, whereas in realistic situations they would indeed be algebraic sums.) One hopes that the brief comments below will give some of the flavor.

Cash from Operations: This is the most basic category,[34] since it represents what the company is all about. It begins with the earnings already calculated on the Income Statement, then corrects for non-cash expenses like depreciation and amortization,[35] and also makes adjustments for net decreases in accounts payable and net increases in accounts receivable. It is comforting to investors to know that the cash is available to provide new inventories and other operations.

Cash from Investing: The most important items here are capital expenditures to acquire new assets (or upgrade old ones) to facilitate the ongoing business. These are recorded as negative (cash outflows) since they consume cash. Such items normally require Board of Directors approval (Chapter 13). Selling old equipment would correspondingly be an example of cash inflow (+).

Cash from Financing: Yes, as the reader has likely suspected, this cash source is (ahem) the borrowing of money. But a critical source of cash nonetheless for companies with some temporary problems. And—note carefully—an absolutely essential cash resource for rapidly growing companies on their way to great success, so one should not look down one's nose at borrowing *per se*.

---

[34]Most accountants would say "cash *flow* from operations," but the "flow" hardly seems necessary, given the "from." This may be similar to those who talk about the velocity of a car—even though the four-syllable word *velocity* is wrong and the one-syllable word *speed* is correct. We should be glad that at least words like "flow," more familiar to engineers, are being used by accountants (instead of the six-syllable and more jabberwockian "financial position").

[35]See, in this connection, the rant about EBITDA in §8.4.

Utility Of The Cash Flow Statement

The Cash Flow Statement seems not to have the sizzle of its two cousins. Who can recall it ever being mentioned in movies, or even books (meaning fiction, of course)? It does smear over a lot of important details—just referring, for example, to Net Income without any hint of the perpetual tug-of-war between revenue and expenses. Nor does it have the historic range or *cachet* of the Balance Sheet, whose depreciation periods can last decades.[36] Still, one can tell a great deal from a Cash Flow Statement, including whether the company has the cash flow to **(a)** make wise capital expenditures, **(b)** meet all debt obligations without crippling growth, and **(c)** pay dividends at recent rates. Pretty important stuff.[37]

It is also worth noting that the term "cash flow problem" should not be used as an excuse or a euphemism for a "revenue shortfall problem" or a "bloated expenses problem" or an "inefficient operations problem" or for any other business problem.

Definition of "Cash Flow" Problem
Simply and only a cash-timing problem. There is a delay between *cash in* and *cash out*.

Remember: A severe cash flow problem can cause company bankruptcy, despite profitability!

A simple example of a cash-timing problem is shown in Fig 8.15.

---

[36] One old U.S. company—notable more for its didactic value than its recent financial performance—had some impressive assets. It was in the railway business and it faithfully listed all its expensive railways and costly tunnels on its balance sheet, with long or near-infinite depreciation periods, as permitted by GAAP. The only trouble was this: Most of its railways and tunnels had not been used for many years, and there was no realistic expectation that they ever would be again in the future. Talk about "tangible" assets! Steel rails and hard rock. But completely irrelevant to the business. One hopes that the banks that lent this company cash are now recouping their losses by lending to knowledge-based businesses, complete with their less tangible assets!

[37] In principle, however, a Cash Flow Statement can be constructed if the Income Statement and the Balance Sheet are available for this year and last. In any case, an old saying about financial reporting goes like this: "Earnings are an opinion, but cash is a fact." Accountants and corporate executives can be quite creative at managing earnings but it's difficult to fake cash.

Figure 8.15: A Simple Cash Flow Problem (A Matter of Timing).

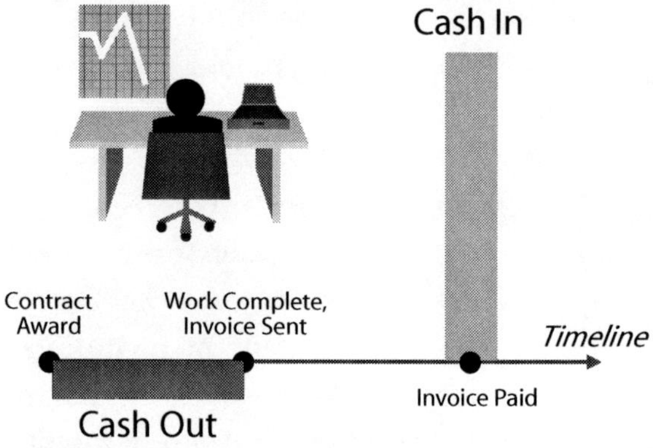

Cash In — Contract Award; Work Complete, Invoice Sent; Invoice Paid — *Timeline*
Cash Out

Notes To The Financial Statements

A basic-but-complete set of the three financial statements (usually at year-end) has been the focus of most of this chapter: Income Statement, Balance Sheet and Cash Flow Statement. However, this still misses an important component—the *Notes* to these statements. To think that the financial goings on of any organization could be summarized adequately by listing of a few numbers and doing some additions and subtractions is patently naive.

The Notes go some distance in filling in some of the missing information and interpretations. Some of these Notes will contain further little accounting statements (usually called *schedules*). Many of these Notes are mandated by GAAP.

Revealing Ratios

Engineers are very familiar with the notion of dividing one number by another number that has the same dimensions, thus producing a *ratio*.[38]

Ratio lovers can have a field day with accounting statements because almost every number has the same units—dollars. In principle, one can divide almost any number on any statement by any other number on

---

[38] If the two numbers have the same dimensions (e.g., length) but are expressed in different units, a simple correction factor (again a ratio, this time a ratio of same-dimensional *units*) is needed. Some writers speak of dividing two numbers of *different* dimensions as a ratio (as in the "mass-to-area ratio"), but this is clearly incorrect; the word they are missing is *quotient*.

any statement and thereby produce a ratio. Accountants are certainly excited; they don't get to divide all that often. The question is: Are all these ratios of any use? The answer is: Quite a few of them, but not all.

If one simply made dimensionless all the entries in an Income Statement by dividing them all by *revenue,* and then compared the resulting dimensionless Income Statement with the similarly dimensionless Income Statement from one's competitors—or even compared it with the dimensionless Income Statement from *one's own company from last year*—many useful inferences could be made. Ditto for the Balance Sheet with everything made dimensionless by *total assets,* or the Cash Flow Statement with everything made dimensionless by *Net Income (Earnings).*

The number of useful possibilities is in the dozens, and many of these have been given pet names[39] and interpreted in ways to help make management or investment decisions. Space does not permit a complete examination of all these accounting ratios here. Most accounting books have a long list of such ratios, with accompanying interpretations.

8.7 MANAGERIAL ACCOUNTING

The distinction between managerial accounting and financial accounting was introduced earlier in this chapter (§8.3). In particular, it was stated that "a mountain of [managerial accounting] financial data underlies" the preparation of a company's (outward facing) financial statements. The patient reader may well ask: Why has so much time been spent on explaining financial statements if these are merely the tip of the iceberg? Why not a larger emphasis on managerial accounting instead?

---

[39] The author's personal favorite is the *acid test* (AT), which, as its name suggests, is a rigorous check—in this case on the immediate cash position. It is defined as follows: AT = (STA − Inv)/STL, where STA and STL are short-term assets and short-term liabilities, respectively, and Inv = inventories. It asks the question: If you had no inventory, or if your inventory were to become worthless, do you have the cash and other short-term (but non-inventory) assets to meet your short-term liabilities—and also purchase new inventory? The idea is that one should have AT > 1. If one's acid test is just unity, all one's incremental working capital must be spent on inventory.

The answer is twofold: **(a)** This is not a book on accounting (it is a book to make engineers think about their career possibilities as managers); and **(b)** while financial statements are, in the barebones forms considered herein, relatively common (as constrained by GAAP) across the business spectrum, management accounting presentations have as many varieties as there are companies. Therefore we shall confine ourselves to a few ideas that are relatively common and that are likely to be of interest to engineers.

Breakeven

Much of managerial accounting centers on *cost accounting*. The simplest cost analysis, which the reader has likely heard of already, is *breakeven analysis*. Here one splits the total cost of making and selling units into a *fixed cost* c_f (which is independent of the number of units sold, n) and the variable cost, $c_v n$, which is proportional to the number of units sold. We don't even distinguish between the number made and the number sold, which would lead to inventory. Thus the total cost is

$$c(n) = c_f + c_v n$$

If the price per unit is set at $\$p$, then the revenue R is $R = pn$, leading to a profit[40] π given by

$$\pi = R - c = (p - c_v)n - c_f$$

The number of units that have to be sold to achieve *breakeven* is found by setting $\pi = 0$:

$$n_{be} = c_f / (p - c_v)$$

This permits an alternate expression for profit: $\pi = (p - c_v)(n - n_{be})$.

---

[40] It may seem almost a sacrilege to some, but business people often use π as a symbol for profit. They spend more time thinking about profit than they do about the volumes of perfect spheres.

Activity-based Costing

A business can be thought of as a process involving thousands of activities, and with each of these is associated a cost. The more precisely these costs can be assigned and rolled up, the more precisely one can manage them. The general approach is to choose the most obvious *cost component* (which managerial accountants tend to call a *cost driver*) that can be directly associated with a particular activity and then assign all other associated costs of the activity to a great grab-bag called *overhead*. Here's an example:

> An employee drives 3 km to purchase an item suddenly needed to complete the manufacture of a product. The most direct cost is the purchase price of the item. Yet, there are many other costs—employee time, transportation costs, purchasing department costs, etc. These latter costs are temporarily ignored and a picked up later as "indirect costs" or "overhead."
>
> This bit of overhead, which is lumped with thousands of other bits of overhead throughout the year, eventually piles up as the OVERHEAD for the year, which is then, the following year, smeared dollar-for-dollar over every direct cost like thick peanut butter. Thus, the cost of the item, say $100, is recorded, after a 150% OVERHEAD charge, as $250.

This is a weak approach to costing, because the same overhead rate is used for everything. Management will not get an accurate view of many actual cost sources and this will lead to poor management decisions.

The ABC (Activity-Based Costing) system is more detailed and thus an improvement. We identify a group of m activities whose costs $(c_i, i = 1, \ldots, m)$ we wish to represent as accurately as possible. To do this we must further identify n cost *drivers* (components) $(d_j, j = 1, \ldots, n)$. Then each cost is a linear combination of the contributions from each cost driver, with coefficients k_{ij}:

Or, using matrix notation, $c=Kd$, where $K = \{k_{ij}\}$, etc.

$$c_i = \sum_{j=1}^{n} k_{ij} d_j$$

This model leads to better decisions about whether a particular activity cost coefficient k_{ij} has really been minimized, whether the activities have been carried out in a cost-effective manner, and whether a particular activity should be done at all.

There will always be an irreducible minimum of unassignable costs left over; these can, with reluctance, be treated as overhead—and then spread (this time very thinly) on everything else as an indirect cost: $c_{i\delta} = (1 + OH)c_i$, where OH is now a much smaller percentage than before.

CHAPTER 9

Innovation

Adapt or Die

The word *innovation* is used at every turn nowadays and with good reason. It is difficult to come up with anything bad to say about it. We appreciate it, want to encourage it, and respect those who do it (and we're not just talking about the technical community here). Why is this? In part, it must be the *newness* represented (*nova* is Latin for *new*). The *–ation* suffix refers to a process—something resulting from action. Finally, the *in–* prefix indicates that something is being *injected*. So, based on the most elementary analysis, we realize that innovation is the process of injecting something new into something. We shall present a more complete definition—the one we shall use in this book—in §9.1.

Many things may be new. The question is: Are they also better? How do we define "better"? Here we return to the definition proposed for technology in §5.4: "any man-made thing that one can use to assist one in doing something one wishes to do," and we note, once again, that this is a very general definition that encompasses much more than engineering or R&D. By these lights, new things are also better things if they are more helpful in doing something we want to do, meaning that they are cheaper, or more efficient, or more enjoyable, or better in some other way. Ultimately, "better" is decided (in a free market economy) by the market; that is, people (either as individuals or in a decision role in organizations) will decide to adopt it or buy it.[1]

Some may find this straightforward definition troubling, but what is the alternative? Large buildings filled with highly paid, slow-acting bureaucrats? Why are they in a better position to decide what is better for individuals or companies than the individuals or companies themselves? Perhaps in some cases, particularly highly complex cases, they are. If the environment is not only a free-market economy but also a democracy, the citizens are free to vote against new technologies being used for purposes not in the public interest.

9.1 WHAT IS INNOVATION?

We could stay with "helpful new technology" as our definition of innovation and not be far off the mark, but the management and business flavor of this book will benefit from a more fulsome definition. To the author's taste, the best complete definition has been developed by the Conference Board of Canada (see accompanying box). The newness is there; the concept of process is included; and the "thing" the innovation is being "sent into" is nothing less than the whole economy!

> **Innovation (Definition):**
> The application of knowledge to create new or improved products, processes, and services that add value in the marketplace. While innovation is difficult to measure, its key outcome is productivity
> —defined as getting more output per input—
> and its ultimate payoff is income and quality of life.
> —*Conference Board of Canada*

---

[1] Not infrequently, one person or group may find a new technology "helpful," while another person or group may find it distinctly unhelpful. It is rare when everyone welcomes an innovation with equal enthusiasm. From the point of view of our definition, the new technology is an innovation if it's helpful to the person or group that wishes to use it. The existence of naysayers, and whether or not they eventually win their point, is philosophy or politics, and far beyond the subject here.

Chapter 9: Innovation

It is not difficult to see why so many people are fond of innovation: by definition, it leads to higher incomes, better standards of living, and enhanced quality of life.[2]

The clause saying that innovation is difficult to measure should be interpreted to mean **(a)** that it is difficult to measure the degree to which it is *currently* happening (in a company, a laboratory, or a country), and **(b)** that it is even more difficult to *predict* how much it is going to happen some time in the future, based on choices of projects, people, technical areas, etc. (We shall return to this difficulty later, in §9.3.) It is more straightforward to measure how much it has already happened; one need only use the market test (are people buying it and using it?). The kind of economic statistics kept by the advanced economies are sufficient to reveal this information.

Role Of Innovation In The Economy

The above definition of innovation places a heavy burden on innovation—nothing less than wealth creation and improved standards of living. Fig 9.1 shows that the so-called "factor inputs" of land, labor and capital are magnified in their effects by using increasingly sophisticated technology.

Figure 9.1: The Lubricating Effect of Innovations on the Economy.

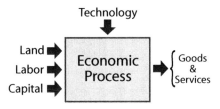

An electrical engineer may use the metaphor of *amplification*; a chemical engineer may prefer to say technology is a *catalyst*. Whatever

---

[2]The author is not so naïve as to be unaware of the fact that most technologies can be used either for good or ill—however "good" or "ill" might be defined. However, so as not to become sidetracked into such issues, and despite their great importance, we shall assume that some political process (hopefully including the free market) has rendered its verdict on each technology in question.

the analogy, when people who want to do something[3] are given powerful assistance through new technology, they can do many more things.

One sweeping example is agriculture. In the most primitive cultures, land had to be ploughed and irrigated by hand. Virtually everyone was involved, all their waking hours, every day of the year, in securing enough food. To these people, the idea of using an ox to help with the plowing was an innovation of the greatest significance. The latest equipment for cultivating, seeding, watering and harvesting is the technological progeny of the ox innovation.

Not Everyone Benefits Immediately

When innovations occur, not everyone stands to gain immediately. Here are three groups of individuals who, while likely generally supportive of innovation, may decry a particular innovation that adversely affects *them*, at least temporarily. Suppose that Company A innovates, while its competitors, Companies B–Z, do not.

> It is inescapable that increased productivity—described as innovation's key outcome in the above definition—means that the same number of people can produce more, which often in practice (since there is usually not instantly more to be done) means that what there is to do can be done by fewer people. Some *employees* may be made redundant by some innovations. However, innovations that lead to new products and services—and even new companies, as in Chapter 12—provide more than an antidote to these effects. Sometimes, the folks made redundant at Company A are relocated to other functions in the company, to the mutual benefit of both them and Company A; other times they are laid off. Such people, at such times, are not well pleased by this particular innovation. Still, to argue against innovation on the grounds that some people may lose their jobs would inevitably lead to the *reductio ad absurdum* Luddite[4] position.

---

[3]Referring to the definition of technology given in §5.4.
[4]Named after General Ned Ludd who, with his little Army of Redressers, destroyed the new machines that textile companies began to use in 1811 England. In a three-week period, over two hundred stocking frames were destroyed. Nottingham authorities had to enroll four hundred special constables to protect the factories. (New jobs, all, apparently!) The "Luddite" epithet is used nowadays to refer to someone who is against all innovation on the grounds that jobs may be lost.

Chapter 9: Innovation

The unfortunate individuals outplaced by innovation should not be forgotten. They can upgrade themselves; they can get career counseling and conduct job searches and, if necessary, they can get further training. Some government support for these replacement activities is completely ethical, since a modest fraction of the economic benefits derived from the GNI (Gross National Innovation) would make a big difference to these individuals and help to complete the cycle of benefit to all.

A second group of individuals who may suffer from the consequences of an innovation in Company A are all those associated with A's *competitors*, namely Companies B–Z (employees, suppliers, etc.). Their fortunes, absent any innovative reaction of their own, will likely wane as their customers move to purchase instead from Company A, either for new functionality, or for lower prices—or sometimes just because it is new. Companies B–Z must react, and their only long-term strategy is their own innovation.

A third group adversely affected by Company A's innovation will be the stockholders (i.e., the owners) of Companies B–Z. Some of our friends and neighbors, who do not comprehend the modern economy, aver that they are not especially worried about those rich people who own stocks. They do not seem to realize that the age of the so-called robber barons is over, and that the ownership of companies (through the stock market) has become much more widespread. Their indifferent attitude towards a decline in stock prices is replaced by disquiet when it is pointed out to them that their grandmother's pension[5] may be based largely on corporate securities.[6] Stockholders, like employees and all others involved in a dynamic, innovative (that is to say, successful and growing)

---

[5] Indeed, some of their own pensions may become more dependent on the stock market. President George W. Bush is currently campaigning to replace at least some of the social security system with a more private-equities-based alternative. Might the increase in the demand for equities depress their returns enough to undermine the scheme? (The laws of supply and demand apply to stocks also, and the supply of earnings will not have increased.)

[6] This week, the sentencing hearing for Bernie Ebbers was completed. Bernie had famously declared earlier that "God is the CEO of WorldCom," but Judge Barbara Jones decided that the CEO had in fact been Bernie. One of the most poignant victim-impact statements was made by a woman who explained how Bernie's US$11B accounting fraud had cost her and numberless others their entire pensions. Judge Jones may well have had that woman's words in mind when she sentenced Bernie to 25 years in the slammer—perhaps the longest sentence ever recorded for a white-collar crime. Bernie's madcap spree of endless acquisitions and debt-hiding accounting does not qualify as innovation by the definitions used here.

economy, have to take whatever corrective portfolio actions that seem best to them.

The above illustrations of innovations, companies, and the economy are not limited to the so-called tech companies. They apply to all companies (and to all organizations, whether for-profit or not) because, as we shall presently discuss, innovations come in many more flavors than "technical." To think otherwise would just be vocophilia on the part of applied scientists and engineers.

Figure 9.2: Innovation vis-à-vis Invention.

Much More Than Invention

The word innovation is sometimes confused with a very similar word, invention. While they do share some characteristics—specifically, the emphasis on newness—they are in other respects almost polar opposites (Fig 9.2).

A technical invention is the creative work of one person, or at most a very small group.[7] It is simply a new idea, some novel approach to something. It may be some diagrams in a filing cabinet. There may be some crude prototype as a proof of principle. Non-technical inventions are similar; they may start with nothing more than a creative conversation at the water cooler.

Innovation starts with invention, but goes much further. Before an invention can become an innovation, many people (perhaps thousands) have to add their special talents, and many more peo-

---

[7]Patent attorneys might like to restrict "invention" to those that are patented, but in everyday language there are many inventions that are never patented. We shall discuss intellectual property in §10.3.

ple (perhaps millions) have to adopt it as their own.[8] An innovation must "add value in the marketplace," which implies that the concept is not just a gleam in an inventor's eye, or even a careful description on a patent application; it has to be fully developed by a complete business team and successfully sold to enough customers that it becomes a worthwhile business and adds value to the economy. There are hundreds of interesting inventions for every truly successful innovation.

We are fortunate to live in an age with so many clever inventions and exciting innovations. At least in the developed world, we have come a long way from the suffocating sameness and the stifling noncreativity that our fellow humans experienced in the centuries known as the Dark Ages.[9]

Innovation Is Everyone's Business

Everyone should be involved in innovation, and in two ways. First, as mentioned above, true innovation needs the coordinated efforts of many diverse skills and specialties. Something doesn't go from being a light bulb brightening in someone's head (invention) to an economically significant entity that has been developed, marketed, bought and used by a noteworthy segment of society at large.

Second, everyone should try to be involved in the seminal creativity of the nuggets of invention themselves. As engineers, we tend to think of the titanic strides that have been made in, say, computers and information technology in the last quarter century, but there is much more. The building of the Hubble space telescope, revealing new wonders in our universe; the decoding of the human

---

[8]There can be some debate about how far this process has to go in order to finally reach "innovation" status. Some say it has to permeate the whole economy. Others feel that it need only be used meaningfully in the originating organization. From the definition we are using here, there must be some nontrivial business (or organizational) consequence to qualify. This will then be reflected in outputs, including customers.

[9]William Manchester, in *A World Lit Only By Fire* (Little, Brown & Co, 1992), writes that "Any innovation was inconceivable; to suggest the possibility of one would have invited suspicion, and because the accused were guilty until they had proven themselves innocent by surviving impossible ordeals—by fire, water, or combat—to be suspect was to be doomed."

genome, promising startling new advances in medical science; the long strides in brain anatomy and physiology, providing new understanding of our true souls as *homo sapiens*; the list in science and engineering is breathtaking. All of these are profound innovations.

Still, we must broaden our sights even more. In the sense used herein, innovation is not limited to mathematics, science, engineering and technology. Indeed, we have already said that innovation is not possible without many more human skills and specialties being involved.[10] Here are some examples of innovations that are not technical but that can have significance for the success of an innovation (from an infinite list): new market niches; new ways to motivate employees; opening doors to new clients in a new way; new alliances (suppliers, universities, other companies, etc.); new product for an existing market; new ways to control costs; new uses of technology for administrative functions; hiring appropriate employees; new financial sources; establishing customer needs in a new way; new ways to compensate employees; new market for an existing product; new ways to get ideas from employees; process innovation in manufacturing (new ways of *making* things); process innovation in services (new ways of *doing* things).

Innovation is an Attitude.

Yes, innovation is an attitude, and this attitude does run somewhat counter to the aversion to change typical of most humans (§7.2). Certainly if one spends *all* one's time questioning *everything* that's going on, this may be a symptom more of an attention deficit disorder than creativity. On the other hand, if one can go a month (or

---

[10] We must remember that the billions it cost to create a successful Hubble telescope was not all spent on engineers and astronomers. The people who engineers don't naturally think about—but who were just as essential (think about that!) to its success—is the army of non-engineers and non-scientists that were intrinsic to all the underlying organizations involved. They were *just as essential* in the sense that, without them, Hubble would still be just a great technical idea. This should not be taken to mean that bureaucracies cannot become too large or too inefficient—only that engineers can't do it all themselves. There has to be a management component for any engineering project to be successful.

even a year) without coming up with a single good idea for improving how things are done, the diagnosis may be incipient cerebral sclerosis. Some people are just more creative than others; some tend never to question the *status quo* while others take nothing as absolute and final; all functioning individuals, however, should be, or should try to be, part of the unending creative process.

9.2 INNOVATION IS LONG, HARD WORK

It has been stressed above that innovation is hardly limited to science and technology. Still, since this book is written for readers with an engineering background—and more specifically, this Part II is intended for young engineers just nicely getting into management—we shall focus more tightly now on technical innovation. Much will also apply more generally to innovation in the larger sense.

The Timeframe For Innovation Can Be Decades

Or even centuries, if one includes the origins of all the mathematics and science sometimes involved. Here one should glance again at the Long Path diagram (Fig 4.1).

However, Fig 4.1 was drawn to emphasize the maths, science and engineering relationships, while here we are getting to the final stages of successful innovation. These are best carried out by for-profit corporations. This may seem like a political statement, but one has to go very far left in the political spectrum (basically off the coast altogether and into the Pacific Ocean) to get any serious disagreement on this point.[11] Figure 9.3 shows some of the typical areas of activity within the modern corporation. Not every corporation will have every one of these components; but large corporations will have all of these and many more.

---

[11] This is not to suggest that all private corporations are perfect; far from it. They have to be made to operate within a set of rules, societally derived. But the modern corporation, for what it is intended to do, is the best vehicle for the commercialization of innovations—and is itself an innovation of historic importance, as anyone observing developing (or previously centralized) countries trying to create such a system will quickly note.

Figure 9.3: The Corporation and Some Common Components

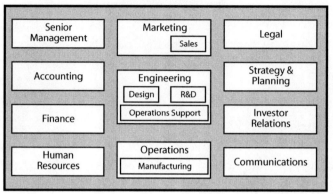

Note that engineering is just one of many components, although an important one. Even for high-tech companies with their own R&D departments, those who have studied technology innovation reckon that the R&D itself is at most 15–25% of the whole process. Figure 9.3 also shows, once again, that the innovation doesn't all happen in the R&D group, or even in engineering; every trade, specialty and profession should continually be renewing and improving itself. The total company team has at least dozens and perhaps thousands of members; how they interact as a team can have great intrinsic value.[12] All these team systems and processes can themselves continually benefit from innovative approaches and improvements.

Peter Drucker On Innovation

As Peter Drucker,[13] the most important writer of business advice in history, has observed, "There are *occasional* flashes of genius." This is the stuff of movies; it is most welcome when it arrives but cannot be relied upon as a strategy. We all have appreciated the creative

---

[13]See Intellectual Capital in Chapter 10 and some examples of the opposite in Chapter 6, especially §6.4.
[13]Peter Drucker, *Innovation and Entrepreneurship*, Harper Business, 1993.

genius of a few people, but we can't leave innovation entirely to them. Innovation is hard work, rarely serendipity.

Drucker goes on to say that "Most innovations result from a conscious, purposeful search for innovation opportunities, which are found in only a very few situations." (See accompanying two boxes.)

Generic Internally-Driven Opportunities for Innovation
[after Drucker]

- Unexpected occurrences: *Recognize, and exploit, serendipities!*
- Incongruities: *Find the most difficult part—and find a new way to do it!*
- Process needs: *Find the bottleneck—and innovate to expand it!*
- Industry & market changes: *If well handled, these provide enormous opportunities!*

Drucker argues convincingly that brilliant ideas for innovations are rare, unpredictable and seldom successful from a business standpoint.[14]

Generic Externally-Driven Opportunities for Innovation
[after Drucker]

- Demographics: *Long-term market segment!*
- Changes in societal perceptions: *Tuning your message to your customers' meanings!*
- New knowledge: *Respond to discoveries and innovations of others!*

For true innovations, Drucker says, the greatest praise one can bestow on an innovation is "That is obvious!" Moreover, if an innovation does not strive for market leadership from the beginning,

---

[14]To be successful in a business context, innovations must be market driven, preferably sooner—but, otherwise, not much later.

Drucker says, it is not innovative enough. With innovation, as in many other areas of life, timing is very important (Leonardo da Vinci was a bit premature with some of his innovations).

In addition to analyzing opportunities (as in the two boxes above), Drucker advises that proto-innovations (present author's word) should be scanned by both left and right brains (both analysis and synthesis). The idea should be simple and focused, and scalable.[15] Also, don't try to be too clever—ordinary people will be using it!

To complete Drucker's advice, innovations, to be successful, should build on the company's strengths (a.k.a. "core competencies"). A scattergun approach is not likely to be successful. Finally, innovators are not risk-focused or risk-driven; they are opportunity-focused and opportunity-driven. (Recall the risk-reward discussion in §7.6.)

Example—The Six-Hat Process

The following example illustrates how determined one company was to innovate, and how it went about it.

> In the late 1990s, MDS Sciex, a Canadian health science equipment company, decided it was so reliant on technical innovation that it had to become even more proactive about ensuring that this innovation took place to the maximal degree. It decided to use a Six-Hat process—a generalization of brainstorming.
>
> In its simplest form, brainstorming is a group of colleagues trying to come up with some new ideas. There may be a specific problem that needs solving, or it may be just "blue sky" creativity. As we use it here,

---

[15]Engineers should have no trouble understanding the broad concept of *business scalability*, because it is closely allied to the mathematical concept of *linearity*. If a business is linear (author's terminology), then doubling all the enabling financial inputs (shareholder investments and lender loans) should, with appropriate management actions, and after an appropriate time lag for adjustment, lead to a doubling of financial outputs (earnings being the most important of these). Even more exhilarating, microeconomics reveals the importance of *economies of scale* and *economies of scope*, both of which indicate that, if the underlying business is both linear and scalable, the increase in outputs (earnings) should be *more than* proportional to the increase in financial inputs. Small wonder that venture capitalists look for scalability (among many other things), since growth in the young company will be possible and positive, not self-limiting by some nonlinear obstacle. (For more on this topic, see §12.2.)

and as a prelude to the Six-Hat process, *brainstorming* has more structure. It comprises two phases. In Phase I, participants are encouraged to come up with as many good ideas or solutions as they can. The atmosphere is meant to be encouraging, relaxed, friendly and generally as receptive to new ideas as possible. Most especially, it is absolutely *verboten* to criticize any of the ideas—including one's own (by refraining from contributing fully, or on the grounds that one "can't think of anything brilliant," or that it "probably won't work").

In Phase II, the ideas produced by Phase I are examined in a more practical way. Everyone knows that, in the end, some ideas are better than others and that the purpose is creatively to solve the original problem in the best possible way. Doubtless, some ideas that everyone forbore from criticizing in Phase I, now get shot down in (hopefully friendly) flames. This process does produce better ideas than its arch-opposite: doing nothing and leaving matters to providence (Drucker's "occasional flashes of genius," or, worse, deadwater).

The Six-Hat process is a considerable extension of, and a major improvement to, the Two-Phase brainstorming process. A sizeable group, spanning many specialties and areas of expertise, is gathered—and, yes, they all wear hats that are identical except for the color, whose significance is as follows:

Blue Hat: Only one of these. Responsibilities include setting the agenda and watching the time, keeping disciple and focus, and generating the summary and conclusions. In short, Blue Hat acts as Chair.

White Hat: As few as one of these, but possibly more. Responsibilities include identifying existing and missing information, and determining the quality and relevance of the information.

Green Hat: A goodly number of these. Responsibilities include the search for new ideas and possibilities. This is essentially Phase I of brain-storming, except that these folks are really good at it.

Yellow Hat: Responsible for assessing the advantages and benefits of the new ideas.

Red Hat: Responsible for stating feelings and intuition, without justification.[16]

---

[16] The author will not attempt to justify this function but includes it for a complete description.

Black Hat: Responsible for giving cautions and identifying difficulties and problems, preferably of the logical variety.[17]

If any Hat starts to wander into the roll of another Hat, the Blue Hat can ask them to desist. According to the reference source,[18] the Six-Hat process **(a)** harnessed focused thinking within a flexible process, thus appealing to most engineers; **(b)** discounted the belief that creativity is the domain of only selected people; **(c)** generated consensus; and **(d)** depersonalized criticism.

Based on team members' estimates, for every $1 spent on this program, the company got $26.48 back. Even if this return claim is high by an order of magnitude, it is an impressive return. It effectively addressed the reasons that most design and innovation arguments arise, namely: **(a)** presentation of incomplete or erroneous information; **(b)** poor handling of questions (adversarial tone, poor topic focus, interruptions and tangents); **(c)** generalizing and/or exaggerating issues; and **(d)** an inability to generate consensus, identify new ideas, or find acceptable solutions.

The purpose of this example is not to convince everyone to suddenly dedicate themselves to the Six-Hat process, or any other particular process, but to show that innovation is, as Thomas Edison said of genius: "99% perspiration and 1% inspiration." On the subject of innovation, Edison knew whereof he spoke.

9.3 RESEARCH AND DEVELOPMENT

In this section we shall focus on what comes to mind for most engineers when they hear the word "innovation"—namely, applied science research. This focus is bred in the bone, as described in Chapter 2, because almost all engineering schools emphasize applied science research as the highest value and the most sacred calling.

---

[17]Should we be surprised that this hat color is black? We've all known people like this; they aren't particularly creative but they can always criticize the ideas of others. Now such people have a valid role to play in the innovation process!
[18]*Engineering Dimensions* magazine, May–June 2000.

There can be no argument about whether applied scientists—whether trained in an engineering school or transplanted from a science department—contribute enormously to the early stages of technological innovation. These are the creative processes that are responsible at a fundamental level for the richness (and much of the riches) of our society. The only issues raised in this book, written primarily for engineers, and intended to broaden the horizons of those who want a wider view, are these:

- There is much more to engineering, generally, than applied science. On the one hand, a creative and tuned-in applied scientist is indeed a precious collective asset. On the other hand, an engineer who does not become an applied science researcher is neither second-rate nor a failed researcher.

- One of the primary skillsets in engineering is management, with its emphasis on results that are useful, team-generated and that meet goals set by the employing organization. Although most engineers engage in significant management activity (after they graduate; see Chapter 1), only a few percent of these will become power managers at the highest levels. Still, surely these latter young women and men are precious resources also, with much to offer society and their employers.

If this book convinces only a few percent of all the outstanding young engineers to take their potential management careers seriously, it will have met its objective.

The Role Of Universities In Fostering (Technology) Innovation
This subject requires that we examine that ubiquitous acronym: R&D. Based on the disciplines of mathematics and science (Chapter 2), and with some important applied skills added to give a practical focus, the *research* half of the pair is innately familiar to engineers with a modern training. Although in other intellectual or academic

pursuits, research has a somewhat different flavor—for example, in the arts it has more of the "studying and looking things up" essence that one would expect from the basic construction of the word RE-SEARCH—the trait of *originality* is fundamental to engineering research. It is an absolute requirement of all[19] the doctoral engineering theses that the author has ever encountered and it is a highly prized (if not mandatory) attribute of a master's thesis from most engineering schools.

This speaks very well for the future of *technological innovation*, a first cousin (and more likely a first-degree sibling) of *originality*. It is evidently compulsory that successful doctoral students (and also students in most less-advanced graduate programs) innovate! The key question, of course, is this: Once the degree has been awarded, and once the de rigueur full-length, original paper has been published in the peer-reviewed archive journal, what happens? Occasionally, a great deal; sadly, more often nothing. Many academic research supervisors are judged—by their promotion and tenure committees, by their salary review committees and by their colleagues—by their research production, meaning the number of full-length, original papers ... [etc.]. In other words, the culture and the reward system in academia is based on the science paradigm. The seeds of innovation are there but there is no attempt to grow them.

This is far from ideal, since many of our most gifted applied science researchers spend their whole lives in this eunuch-like harem of unconsummated[20] potential innovations. Even worse, its graduates have been trained, over the most mature and creative part of their life thus far, to disconnect *discovery* from *innovation*. After watching both academic and nonacademic research—in universities, corporate R&D centers and government and quasi-gov-

---

[19]Several hundred, from many universities.
[20]This remark does not apply to scientists, whose professional paradigm is indeed to discover how nature works and to publish their findings. The problem is that engineering academics are judged, almost exclusively, by science paradigms, not engineering paradigms.

Chapter 9: Innovation

ernment[21] laboratories over four decades—the author is convinced that the most fundamental problem relating to this group is the following: How can we motivate our most creative and highly trained minds to focus on something of higher value than just seeing their name in print in a prestigious journal? How can we get them interested in true innovation, not just in being judged to be a very bright person? And how can we provide this guidance without squelching the very sense of freedom and independence that is the intellectual precursor to important new ideas?

This blight is most especially pernicious in academia. Universities are, in population, the size of moderate cities, and in their own sense of importance, the size of large cities. Many of the leading universities have existed for many decades, possibly for centuries, and are not receptive to suggestions for change. Yet they[22] are surely responsible, not only for the societal worth of their luminous staff, but also for the precious cargo they receive annually from secondary schools and deliver to the world several years later.

Universities are certainly not accountable for everything their graduates do or don't do, but they do bear a heavy burden of responsibility for the state (using the system-theoretic definition) of their engineering graduates when they leave the ivied halls of their *alma mater*. In the case of engineers, most important for this book, surely universities should inculcate a sense of responsibility for accomplishing more than just something new and publishable.

Which brings us to the second half of R&D. What do university professors mean when they use the term "R&D"? Or, more specifically, since we have just discussed the "R," what do they mean by "D"? This is a variable quantity. Some mean nothing at all very concrete, using R&D as just a facile terminology, one that has a firm purchase with their granting agencies, and that is therefore sprinkled liberally in their proposals for research funds. Others are truly

---

[21]"Quasi-government" means that most of the institutional funding and much of the organizational culture flow from government. Most academic research qualifies under this definition.
[22]As always in this book, we speak only of mathematics, science and engineering.

engineers, choosing to spend their careers developing innovative technologies in their research, helping to mould the next generation of engineers, and very willing (although they are not always sure how) to export their discoveries into society at large.

Technology Transfer From Universities

There is an increasing recognition that, in the words of a recent report,[23] "university research is a national asset; we should manage it better. Universities and researchers must more clearly recognize their responsibilities. Universities and government are ultimately responsible to the taxpayers." These seemingly self-evident truths were not received kindly by some sectors of the university research community.

A few years earlier (in 1995), a joint report from the Natural Sciences and Engineering Research Council and the Conference Board of Canada studied the so-called *culture gap* between universities and the business world by asking each what they felt about the other. Companies apparently felt that "universities don't seem to know about private-sector requirements or how precious our research dollars are." And universities typically said that "companies are largely unfamiliar with our research capabilities or the true nature of research activity." This corresponds to a double gap amounting to a chasm—the Inside-Outside Culture Gap first mentioned in §2.4.

This state of affairs can be discouraging; yet, since universities do about 25% of the national research, it cannot just be ignored. And the good news is that things are getting better, not worse. The granting councils and various centers of excellence are making relationships with the private sector (including cash contributions) mandatory for some kinds of grants and contracts.[24] The central administrations of the research-intensive universities have all set up offices to encourage and facilitate technology transfer.

---

[23] "Public Investment in University Research: Reaping the Benefits," Report of the Expert Panel on the Commercialization of University Research, 1999.
[24] The author stipulates, once again, that this discussion is about academic *engineers* (i.e., applied science professors). The protestations from university *science* departments that they should not have to "follow a corporate agenda" may or may not be valid, but they are not under examination here.

There have always been instances of mutually beneficial technology-transfer relationships between individual professors and individuals in the business world. These were not the results of some longwinded memorandum of understanding between two large organizational entities, but of the common interests and goals of a few individuals on both sides who were able to bridge the culture gap. Notably, the "transfer" went *both* ways; the professor learned a lot about engineering in the real world as well.

At present, it is probably best to be aware that significant opportunities do exist for university-business cooperation and that there are more of these every day. One must be alert to attractive situations. Technology transfer will not become a wholesale output of engineering schools, however, until another generation of professors has lived in a more business-friendly culture and been enticed by a reward system that recognizes technology transfer as a higher priority.

Pre-competitive Research

Applied science research (well before product development) is very risky because it is difficult to foresee the benefits with much certainty. Further, it is quite costly because expensive equipment is often needed and its practitioners are highly educated. This combination of high cost and high risk is a troubling one for private-sector firms who must pay for their research out of their revenue in a competitive situation. It is natural to seek ways to reduce the risk and cost of this research.

This problem is even more pronounced in Canada where companies are fewer and smaller than in the U.S. The Canadian Institute for Advanced Research (CIAR) proposed a framework some years ago in which research costs and risks early in the Long Path could be shared among several companies with some shared technical interests (Fig 9.4). While basic research was so costly and risky that only the government labs and universities (whose basic research is sponsored, in turn, almost entirely by government) could afford it, and while explicit product development would remain the province of individual companies, there were intermediate research activities that companies might wish to pursue together.

Figure 9.4: The Benefits of Identifying "Pre-Competitive" Research.

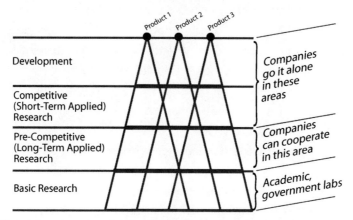

This idea was implemented to some extent and several consortia were created to do pre-competitive research. In practice, these consortia were supported partly by private money and partly by government funds.

Public Sector R&D

There is clearly a role for research and development activity in the public sector. One can debate exactly how large or small that role should be, which is sometimes a matter of political taste. At one extreme—what might be called the "socialist" position—the government must do virtually everything for everyone, from cradle to grave, including research and development.[25] At the other extreme—what might be called the "libertarian" position—the government should look after national defense and not much else. This might be fun for a few weeks, but serious governments, even if they do not choose to

---

[25] In one recent instance, a public-sector union leader responded to a government's new policy of outsourcing $107M of activity to the private sector as "taking $107M out of the economy." Evidently, to that person, no dollar was even in the economy (!) unless spent on government labor—even if the government was doing the spending! This view is not advocated here.

engage in massive wealth redistribution projects, do try to look farther into the future than the next corporate reporting quarter, so as to safeguard the economic security of their jurisdiction. One way of doing so is to engage in carefully chosen R&D activity, particularly in areas that seem strategically crucial to the country.

The criteria for public-sector R&D are (unavoidably) more complex than for private-company R&D. The latter are responsible primarily to their shareholders (owners) and will not reasonably engage in any activity (including R&D activity) that is not in their narrower interests. Government R&D, on the other hand, is (or should be) targeted overall at helping everyone in the country. For example, a country is (and should be) interested in stimulating[26] employment, although, for a corporation, employment is a secondary effect not a primary goal.

It is difficult to argue that the national interest is completely served by the accumulated interests of all current corporate shareholders. In addition to gaps between the sum of all national interests and the sum of all corporate shareholder interests, there are also major discrepancies between the timescales impressed on corporate leaders and the time horizons of importance to national leaders. Companies and countries also have a completely different view of the risk-reward calculus.[27] Modern societies must have an R&D strategy—and, more generally, an innovation strategy—if they are to thrive in the modern world.

There are several modalities for government support of national R&D. These include national laboratories, which can (on the small scale) help companies with modest R&D capability to develop new technologies and which can (on the large scale) study nationally targeted research areas. Long-term (high-risk) enabling and infrastructural technologies can also be pursued in the public sector. As a final

---

[26]There is a fine line between a government R&D program that is intended to stimulate future employment and one that is a social assistance program disguised as an R&D program.
[27]When President Kennedy announced on September 12, 1962 that space R&D was in the national interest, both militarily and economically, and that U.S. astronauts would land on the moon by the end of the decade, he did not say "I really hope some corporation looks into the possibility of space R&D because it might be important in the future."

example of an important species of government R&D program, one can cite R&D tax credits, which help companies to climb over the risky hump of new technologies, yet leaves the picking of winners and losers not to government bureaucrats but to the market itself.

Measuring R&D Impact

Research and development are themselves difficult tasks; they require supple, highly trained minds to perform. Even more difficult, however, is to choose what topics to research and which products (in the most general sense) to develop. Note that we are not talking about mathematics or science research here—although those subjects may be useful along the way; "fundamental" or "curiosity-based" research can rarely be afforded by most companies or their laboratories. We are also not thinking about most academic applied-science R&D because the aim of the "R" is usually just to write a paper and there is rarely much "D."

Still, even with these exclusions, a multitude of possibilities are usually presented to company management, far more than can be afforded. Given that ten years from first research to final commercialization is not an unusual time frame, making wise R&D choices is evidently an expensive and risky game. But play it we must, or we will fall behind. We want maximum commercial impact per dollar spent.

Figure 9.5: Predicting R&D Impact is Very Difficult.

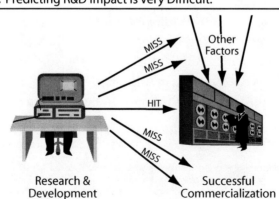

Chapter 9: Innovation

There are two major sources of difficulty in predicting R&D impact (Fig 9.5). The first is the prediction of events far into the future; there will be many misses for every hit. The second is that there are many key activities—other than R&D—that have to be successfully executed[28] to enable successful commercialization; how does one assign credit among the many team activities for the success? In fact, it is even difficult *in retrospect* to decide on R&D impact and to apportion the deserved recognition among the many areas of expertise needed to achieve success, much less predict it years in advance.

Terminology that has been found helpful in discussing R&D impact is shown in Fig 9.6. Note the use of the word "impact." Writing a paper or report, or a researcher getting paid, may be "results," but they are not "impacts." The "inputs" and "outputs" answer the question What. The "activities" answer the questions What, How, Who, and When. The "impact" answers the question Why.

Figure 9.6: Helpful Terminology.

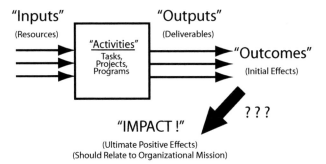

There are many reasons to evaluate R&D impact: the lab manager wants to justify his R&D costs and to choose between new projects; the human resources department wants to evaluate indi-

---

[28]For example, one can forgive the marketing and sales staff for thinking that they have a lot to do with commercial success. The finance folks observe that, starved of funds, the place would shut down overnight. Managers know that it was their successful leadership that made good things happen. The list goes on. (And they are all correct.)

vidual researchers and to design skills development programs; and senior management and the board of directors want to fine-tune company strategy.

It is difficult to find a simple proxy for research and development impact. Universities try to use the science paradigm (number of quality publications), even for engineering research. Other organizations attempt to use patents as an indication, even though most patents never get used and many important products (e.g., software) have never been patented. The pressure is enormous to use some counting technique where **(a)** the count is simple, and **(b)** all the evaluative work has been done by somebody else (reviewers, editors, patent attorneys, etc.). Unfortunately, these proxies are usually poor because predicting and evaluating R&D impact involves a good deal of qualitative judgment and requires hard work. They may range from client surveys, peer reviews and case studies (relatively straightforward approaches) to cost-benefit analyses, Monte Carlo simulations and econometrics (more complex approaches).

9.4 PATTERNS OF INNOVATION

To managers, one of the most frustrating characteristics of the innovation process is its intrinsically unpredictable nature. Progress is highly unsteady, especially at the earlier, more-R-than-D end of the progression. This leads to two fundamental difficulties that make research (and even R&D) managers tear their hair out in overseeing their very-human research team members. First, progress is difficult to measure (see §9.3) and especially difficult to predict. (It is also only human nature to exploit this fact to explain all matter of nonperformance by researchers.) And second, how does one give one's R&D staff the joy and freedom they need to become naturally innovative, while at the same time guide them towards overall team objectives and align them with organizational goals?

These and related anthropocentric issues are more nearly the focus of the next chapter and will be considered there. We can here,

Chapter 9: Innovation

however, take a look at some of the non-psychological aspects of the innovative process that have been established over many years and that, when understood, help to make innovation a somewhat less risky business.

The S-Curve

Many readers will already be familiar with the S-Curve—so called for reasons that will immediately become obvious—and they may wish to take a pass on this section. However, the underlying concept is so important and of such widespread relevance that it is not possible to leave it entirely unmentioned.

Figure 9.7 shows a truth that has many interpretations and diverse applications: The cumulative reward enjoyed as effort is first expended is relatively small; then the reward grows as additional effort is expended, based on the previously furnished foundation; and finally, the additional reward eventually weakens, despite the additional reward expended, a behavior well known as the *law of diminishing returns*.

Figure 9.7: The Famous S-Curve (and its first-degree relative, the hill-curve).

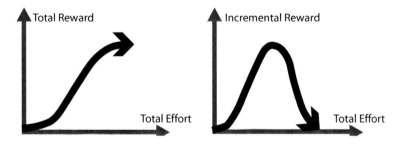

Anyone with training in calculus (or anyone who sees the pattern in Fig 9.7) will notice that the S-curve and the hill-curve are saying the same[29] thing: the hill-curve is the derivative of the S-curve and the S-curve is the integral of the hill-curve.

---

[29]As a mathematical nicety, the S-curve also says that there is no initial reward without initial effort.

The abscissa in Fig 9.7 can be replaced by "time," "manpower" (e.g., person-years), "total financial expenditure," or any other measure of effort that increases monotonically as time goes on. The ordinate can be replaced by "state of knowledge," "state of sophistication," "readiness for product development," "readiness for manufacture," "market readiness," or, for an entire innovation process, "product sales" (compare the product lifetime graph in Fig 5.10).

Generations Of Technology

Figure 9.7 is a very simplified version of events, not just in the trivial sense that it does not purport to represent accurate data for any actual process, but more consequentially because it shows only the lifetime of a particular effort-reward incident. In practice, with innovations, the curve has endless generations, each succeeding the previous (in all senses of the word "succeed").

Figure 9.8: A Simple Technology Family Tree.

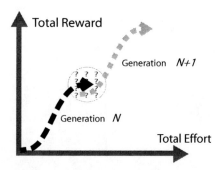

This idea is shown pictorially in Fig 9.8. Each generation[30] of technology—stimulated initially by a bright idea, but then surviving the long gamut of challenges to eventually survive as much more than merely a bright idea—builds upon its previous technology generation to achieve ever higher levels of performance (however performance is measured in each case).

---

[30]We are speaking here about innovation in the most general sense (as defined in §9.1), not just in the more narrow engineering and scientific sense.

Chapter 9: Innovation

As apparently true for humans as well, there is no clear-cut point that defines the end of the previous, and the beginning of the next, generation. This gray area is also shown in Fig 9.8.

The Learning Curve

Another application of Fig 9.7—in addition to technology generations, and its close relative, the product lifetime curve—is the celebrated "learning curve," which applies to individuals, projects, companies, or any other learning entity. This is shown in Fig 9.9 in two versions (distinctions without a difference) because some refer to how much has been learned and others refer to how much has yet to be learned.

Figure 9.9: The Learning Curve (per topic) (two versions).

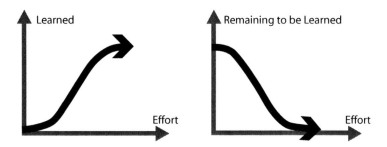

Either way, learning is long and expensive but indispensable to the overall innovation process.

Process Innovation Vis-à-Vis Product Innovation

Utterback[31] and others have pointed out that *product innovation* and *process innovation* tend to take place in a particular sequence. They have noted through the use of many examples that a product tends to be more mature in its innovation lifetime sooner than the process ultimately used to manufacture it. This makes sense: why perfect a manufacturing process for a product that has not yet reached its dominant design?

---

[31] See, for example, JM Utterback, *Mastering the Dynamics of Innovation*, Harvard Business School Press, 1994.

We earlier noted (Fig 5.14) that the distinction between product and process is more contextual than intrinsic: my product is part of my customer's process and my supplier's product is part of my process. Therefore, we can think of a *value chain*[32] in which Company N makes a product (Product N) that is sold to Company $N+1$ which in turn uses it to make Product $N+1$, and so on. As shown in Fig 9.10, serious innovation for an optimal process to manufacture Product N is not economically viable until the design of Product $N+1$ has stabilized (through its own innovation process) to the *dominant design*. This "optimal process to manufacture Product N" will involve its own innovation, often including a (new) *enabling technology*. This process continues along the value chain, with the order of innovation completion being opposite to the flow of goods.

Figure 9.10: Bursts of Innovation Progress Up the Value Chain.

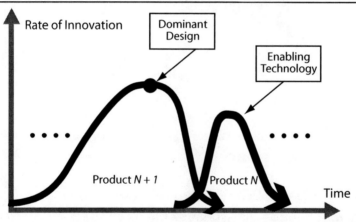

9.5 INNOVATION FRAMEWORKS

By now the reader will surely be convinced that innovation is a very complex process, one that can hardly be left to accident. Even the initial spark of inventiveness (see §9.1) on which true innova-

---

[32]Economists would speak of the *value chain*; business people might more likely refer to the *supply chain*.

tion rests has spawned volumes on the subject of human creativity, and in particular how to stimulate it. Yet, as we have seen, pure creativity is a necessary but far-from-sufficient condition for innovation. Creativity is but the first step in the long, arduous, uncertain process towards true innovation (see definition, §9.1).

Some organizational structures are more conducive to innovation than others, and although one cannot produce innovations by fiat, one can at least build frameworks within which innovative activities are known to flourish. Some of these frameworks will now be considered.

Incubators For Innovation

Incubator is an elegantly descriptive word for an organization that helps startups[33] develop in an accelerated fashion by providing them with a bundle of common services (including physical space, secretarial, bookkeeping, and legal and accounting advice). Also available are opportunities for coaching and mentoring, capital as merited, and networking connections (since an incubator typically has several startups on the premises). Just as newborn chicks need help to grow and thrive, newborn companies can be helped through their initial stage of growth by living initially in an incubator. This improves their probability of success and also speeds up that success; this latter characteristic has led some to use the term *accelerator* in place of incubator.

Incubators come in a variety of species. The most populous group are nonprofit organizations sponsored by governments (national, state, or local) interested in stimulating economic growth and high-value employment in their jurisdictions. It will not be a shocking disclosure that these incubators sometimes suffer from over-bureaucratization and slow times-to-market. Political issues

---

[33]Startups will be discussed at greater length in Chapter 12. Some associate the term incubator with the explosion of information technology companies in the late 1990s, many of which failed when the bubble burst. This was a consequence of the IT frenzy of that period and should bring no discredit to the incubator concept itself. Most incubators do focus on *high-tech* startups; fortunately, this is the group of greatest interest to this book's readers.

and agendas can also undermine their alleged objectives; only the most economically promising startups should be admitted and incubator dwell-times should be rigorously policed.

Some of these incubators are located in close proximity to research-intensive universities. Together, they can form a flourishing partnership and a critical mass. Not only do the startups thrive on the technology-rich environment of the universities, the technology areas researched within the universities have another important receptor for their new development ideas. The Inside–Outside Culture Gap (§2.4 and §9.3) between the academic cloister and the outside economy is thereby bridged. Indeed, some universities themselves sponsor incubators, clearly a signal that they are on-board with the commercialization, where possible, of their research results.[34] Such arrangements also promote interesting summer positions for faculty, topical material in lecture courses, and job opportunities for young graduates.

There are also *private* incubators, largely of two types. In one of these, the incubator is administered by (or at least partly by) venture capital organizations; these tend to be ruthlessly choosy about whom they allow within the incubator, and they require an equity position in the startups under their wing.[35] These incubators, believing in the maxim that one should invest only in what one knows well, and endeavoring to create critical-mass synergies, tend to stress one type of high technology (e.g., information technology, pharmaceuticals, marine biology).

The second type of private incubator is a largely independent entity created by a company of major size. Christensen[36] and others have shown that very large companies—which still must innovate to survive in the long term, and whose innovations must

---

[34] One hopes that these universities also recognize, in their personnel reward systems, any serious contributions to business development made by faculty members from their engineering schools.
[35] The equity requirement is eminently fair, considering the high risk and range of services provided, and is in any case not different in nature for the equity required by any serious investor in any early-stage company (see Chapter 12).
[36] See, for example, CM Christensen, *The Innovator's Dilemma: When New Technologies Cause Great Firms to Fail*, Harvard Business School Press, 1997.

ideally produce a 10% increase in a $1B annual revenue, not a 10% increase in a $1M one—find it difficult to innovate within their existing large-company culture, expectations and accounting system. Having found a hot potential product, they find it best to spin off a separate unit (with its own culture, expectations and accounting system) to further develop and commercialize the product.

In this last instance—the in-house incubator—the development group does not have as much stimulation from outside, but presumably a large company can provide itself as a positive outside influence. For all other incubators discussed above, the fledgling company is nourished by the other continuous influences in its environment: other startups, more mature companies, and possibly a major university. (Having a major market nearby would not hurt either.)

Many of the above characteristics are summarized in Fig 9.11, which depicts the major helpful interactions between the incubator and its outside neighbors. Since all of these interactions, as intended here, are positive ones, perhaps in that sense Fig 9.11 could be thought of as the ideal incubator.

Figure 9.11: The Omnibus (Ideal?) Incubator for Startups.

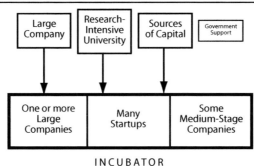

The heart of the incubator is, of course, the several startups who have many things in common, not the least of which is simple survival. Not shown in Fig 9.11 is the coffee room (or cafeteria) shared by the personnel from these startups. This is a crucially important institution since this is where much of the intellectual stimulation and cross-pollination occurs.[37]

This meeting place should be large enough to accommodate the other inhabitants of the incubator shown in Fig 9.11, these other inhabitants being as follows:

> First, some companies somewhat further down the road to maturity, who, unlike the startups, are now paying market price for all their services (including rent), but who feel that they still benefit from the incubator environment. These companies may well be ones that were incubated and then stayed on, but they are no longer a financial burden to the incubator. In fact, they provide a net benefit because they have a salutary influence on the startups.
>
> Second, one or more major (billion-dollar-plus) companies may also be represented in the incubator (paying, need it be said, full market price for everything). Obviously, in these cases, the companies proper are elsewhere, but they have chosen to create a more entrepreneurial atmosphere for some of their research and/or development activities and feel that they will benefit from the energy, innovative thinking and just plain excitement of the incubator atmosphere.

To complete the description of Fig 9.11, the advantages of a research-intensive university and sources of capital (for the smaller companies) have already been mentioned. There is also a role for government—for example, land use, tax incentives, and access to technical and other information—but "contributions" of bureaucracy or political correctness will produce, for an ideal incubator, entropy (at best) or failure (at worst).

There are several similarities between the Ideal Incubator of Fig

---

[37]Disclosure of actual proprietary data by an employee of one startup to an employee of another startup is, of course, an absolute NO-NO, and would likely be a violation of one's employment agreement (with the harmed startup), and thus grounds for dismissal *for cause*; if sufficiently egregious, it may also be the basis for civil action against such an employee. This remark also provides a concrete example of the sorts of advantages incubated startups have: a solo startup would not likely have had time to install employment agreements with all its key employees, whereas an incubated startup, with the legal advice ideally available, would likely *have* such employment agreements—with the appropriate intellectual property, confidentiality, non-competition and non-solicitation clauses included.

Chapter 9: Innovation

9.11 and the Pre-Competitive Research Model of Fig 9.4. In practice, the former benefits startups most, while the latter is of greatest interest to established companies. However, combinations of these ideas may be optimal in particular cases.

Technology Clusters

Yet another organizational structure that has sparked interest from the viewpoint of innovation (and other benefits) is the *technology cluster*. This shares some aspects of the incubator (just discussed) but is on a larger scale and is not rooted *primarily* in the need to nurture new companies. Its basis is the well-proven fact that clusters of high-tech organizations benefit from their mutual proximity and do better as a cluster than a similar collection of organizations acting in isolation.

These knowledge-based clusters, the largest of which will take a generation to bring to full maturity, vary in size from a relatively simple *research park* to a whole geographical region (Silicon Valley being among the most famous). Only three features are common: they are knowledge-based; they lie in mutual proximity; and they benefit from key linkages (i.e., the knowledge bases of all members have a substantial degree of overlap). The innovative effusions of any one member have spillover effects to other players in the cluster, and despite all the advances in remote communication, nothing fosters human interaction like face-to-face personal contact—the consequence of physical proximity. These contacts include the informal water-cooler or coffee-bar type (where, once again, proprietary knowledge and intellectual property must be respected) and also the more formal variety under the *aegis* of a professional association, seminar series, organized conference, or similar.

Further, the cluster attracts a local infrastructure (e.g., financial, legal and IT services) that supports the cluster's needs. "Quality of life" opportunities also tend to thrive in the neighborhood of a large cluster; knowledge workers are typically very mobile and will tend to diffuse away from banal (or worse) locations and towards

267

ones that stimulate and nourish their intellectual, cultural and recreational interests. Universities that are research-intensive and involved in industry are also key elements in this kind of clustorial mix; if firing on all cylinders, such a cluster can and often does foment tsunamis of innovation.

CHAPTER 10

Intellectual Capital

Operating System for a Network of Brains

Something just doesn't add up! When we try to figure out how much a corporation is worth, there are competing indications. Moreover, the discrepancies are not just a few percent here or there—which might be explained by the natural differences between unlike methods of estimation—the different estimated company values are sometimes off by a country mile. This conundrum is particularly true of high-tech companies, which raises flags of both interest and concern for our purposes in this book. As we shall see presently, these discrepancies, interests and concerns are closely connected to the chief subject of this chapter: the value of so-called *intellectual capital*.

10.1 WHAT IS THE VALUE OF A CORPORATION? —ACCOUNTANCY'S CURRENT ANSWER

One of the primary indications of a company's value must surely be its *book value*, as methodically figured out by professional accountants after scrutinizing the companies accounting books. In Chapter 8 (§8.5), we noted that the (book) value of a company, according to accountants, is the *shareholders' equity minus any intangible assets*.

For the majority of companies historically, and even for many companies now, the subtraction of the intangible assets was a small corrective factor but hardly the determining aspect of value. For many intents and purposes, the book value was essentially the *shareholders' equity* (assets minus liabilities).

Tangible Assets

Tangible assets were discussed briefly in §8.5. The basic idea is this: The most *tangible* assets are hard, *physical* assets—assets that are literally touchable. These are normally recorded on the Balance Sheet at the value at which they were acquired (purchased), less any accumulated depreciation. Such assets include desks, real estate (land, presumably improved by buildings unless we are speaking of a real estate development company), computers, manufacturing machinery and myriad other categories[1] of physical assets. We note for future reference later in this discussion that there is always one or more *underlying financial transactions* upon which this asset value is based—plus a depreciation schedule that, one hopes, has some strong connection to reality.

Accountancy recognizes a second category of asset as tangible, namely, most types of *financial* asset. The *exemple nonpareil* is *cash*, whose mild shortcomings of ever-shifting valuation are more than compensated for by its complete liquidity. Many other financial assets are also accepted by accountants as tangible, including many types of financial security, such as stocks or bonds, despite the fact that **(a)** these financial assets are not physical, and **(b)** their value can vary quite significantly with time.

Financial assets do have unambiguous transactions associated with them. One can point to these to claim an independent objective value. But financial assets are not at all physical or touchable. The referent transactions are highly legal in nature and in fact one might guess that accountants have historically paid tribute—per-

---

[1] Three examples of these other categories—all of them inventories—were also mentioned in §8.5: raw materials, work in process and finished goods.

Chapter 10: Intellectual Capital

haps in exchange for some reverse concessions[2] to realism—to a closely-allied profession (lawyers) by accepting as tangible a class of asset that clearly is not touchable.

That Intangible Word, "Intangible"

Next we come to the category of intangible assets—a category that, like all classifications defined as a negative, is quite vexed. Thus far in our discussion, an investment in a highly volatile stock, which can change by several percentage points from one day to the next, is treated as a financial (i.e., tangible) asset. What then, pray, are intangible assets? The answer is: Assets whose value is even more nebulous (and thus more subject to debate and argument) than physical and financial assets. Here are two examples:

Goodwill: This is (as we shall shortly see) an example of intellectual capital. Suppose Company A acquires Company B, which has assets A_B, liabilities (mostly debt) D_B, and resulting shareholders' equity $E_B = A_B - D_B$. How is this transaction, once consummated, recorded on Company A's books? Certainly all the assets in A_B are added on the left (asset) side of Company A's Balance Sheet, as are, on the right (liability) side, all the liabilities in D_B. But this creates an imbalance: The difference, namely, $E_B = A_B - D_B$, creates the imbalance. The solution from accountancy is **(a)** to record E_B as a new asset, called *goodwill*, on the left side of Company A's Balance Sheet, and then **(b)** to amortize[3] this asset away over a fairly long period of time. At the time of the transaction, the equity of the acquiring company also goes up by an amount E_B, to keep the Balance Sheet in balance. Evidently, intangible assets are indeed an integral part of accounting; the more modern challenge is to extend the set of such assets that can be so recorded.

Patents: We shall discuss patents later in this chapter as one element of a company's set of intellectual *property*, which is, in turn,

---

[2] Possibly a disinclination to litigate?
[3] The precise time rate of change (negative) is very important to accountants but it's the concept that interests us here.

a subset of intellectual *capital*. For now, we note simply that when a patent is created on a technology, this is recorded as an asset on the balance sheet. So far, so good. The key next questions are: **(a)** At what value is it recorded? And **(b)** How is this asset treated with respect to changes over time? Here, the accounting picture is not very helpful. The patent is recorded at the actual legal cost of creating it—let's say $10K for discussion purposes—which bears no relation at all to its actual value. In addition, sometimes this value ($10K in our example) is amortized over time, sometimes not.

Accountancy's Treatment Of Intangible Assets

First, to summarize: Accountancy's treatment of the value of intangible assets is (to be kind) unhelpful. To be less than kind, accountancy's treatment of intangible assets is abysmal. It will take nothing less than *disruptive innovation* in the accounting profession to improve the situation. Most of the principles of this age-old profession were developed in ancient worlds and were applied several millennia (or at least centuries) before phrases like "intellectual capital" and "intangible asset" became known and important. Most of these elderly principles are still important today, having been slowly developed to provide some certainty—and some protection[4] against the ever-innovative financial scam artists—but the theory of accounting must grow beyond these bounds if it wishes[5] to respond to civilization's needs and level of understanding, just as have all other professional disciplines that are estimable, established and in need of expansion.

We shall presently see that, under effective management, goodwill (and other intangible assets) can be made to grow right along with

---

[4]This is actually to give the modern accounting profession more than it deserves. Most of the recent corporate accounting scandals were perpetrated, not just by ingenious internal crooks overwhelming the eternal scrutiny of auditors, but with the express complicity and creativity of many of the largest accounting firms, who assisted their corporate clients in bilking trusting employees and small investors out of many billions of dollars in exchange for multi-million-dollar consultancy fees.

[5]The more thoughtful members of the accounting profession most emphatically *do* wish to respond, as briefly documented later in this chapter, but the long-term development of the profession along progressive pathways will find stiff competition from the short-term and well-paid calls for "creative" accounting in the more negative sense. Surely the accounting research groups at leading business schools are in a position to lead the charge on this issue?

tangible assets. The forced wasting away (amortization) to zero of goodwill and other similar critical intangibles has more to do with the inability of classical accounting to deal with the valuation of intangible assets than it does with reality. Generally an effective business tries to enhance its customer base—the basis of the word "goodwill"—in which case there is no basis in reality for such assumed shrinking.

When it comes to patents, the story is similar. If the patent is just a legal document collecting dust in a filing cabinet, and no successful effort is being made either to develop the patented technology as a business enterprise or to license this technology to some other company in exchange for royalties, then a superb estimate of the value of this particular intangible asset is $0.00.

One key principle in accountancy—and one that has obvious attraction—is that values be *transaction-based*, meaning this: a willing, experienced, informed, sound-minded *seller* sold the object in question to a willing, experienced, informed, sound-minded *buyer*. Every word in the previous sentence is intended to foreclose the possibility of whining or litigation against one's provider of accounting services. How can one argue against the value to which such a seller and such a buyer have mutually agreed? If the object sold is a very tangible wheelbarrow, the short answer is: one really can't.

Intangible Assets—A Valuation Challenge That Must Be Met

For intangible assets, on the other hand, whose values are extraordinarily sensitive to how they are affected by further intellectual activity, the *transaction values* themselves, even if representative at the instant of transaction, have a highly transitory utility at best, and possibly none[6] at all. The evidence of value provided by actual

---

[6]Often, recorded values of intellectual assets, even at the time of their creation, are bizarre. Patents are an important and common example: If a team of world-class engineers, scientists and technologists devises a new product or process (at a fully loaded cost of $3.5M) and then a patent attorney arranges for the associated patent (at a fully loaded cost of $10K), how can it possibly be that the resulting patent is logged on the books as an intangible asset valued at $10K? If the R&D is separately recorded, this two-and-a-half-order-of-magnitude discrepancy is clearly explained; but, if not, the other explanations offered with straight faces will not be comforting to scientists and engineers—or to anyone else. Perhaps the professions of law and accountancy should admit new members to their club?

transactions is compelling if **(a)** the "transaction" is relevant to the valuation, and if **(b)** the transaction is recent enough to be pertinent. For many—in fact, for most—components of intangible assets (especially intellectual assets) there is no simple transaction at all, either recent or relevant. (It is simply something that the company has been doing.) A healthy profession (for example, engineering) does not throw up its hands in the face of a difficult but highly important calculation; it rolls up its sleeves and, after a few decades at most, it provides the needed way through.

Another principle of accountancy is *objectivity*. While hard to argue against in the abstract, in practice this principle in the absence of any other principles again leads to the reliance on actual transactions and this in turn means that only *transactions that have already happened* are admissible on an official financial statement. We have seen this phrase before. It was noted in the Corollary to The Second Principle of Business Risk in §7.2. Accountants, in their roles as reporters to the tax department, will not get involved in anything that requires financial projections into the future, and presumably most companies would like to keep things that way. It would not strain credulity to predict that the government would like to initiate tax levies based on future prospects as well as on past (actual) financial performance.

Still, this is beside the main point. Companies cannot at present retain accounting services that will take into account future projections of their financial welfare based on the effective exploitation of their intangible assets, nor, indeed, can they even ascertain the theoretical value of these intangible assets. This would appear to be an enormous and obvious market for modern accounting services—of interest to management, lenders and investors, among many others.

A Counter-Argument

Finally, some countervailing points. Consider the diagram shown in Fig 10.1, which shows the continual creation and nourishment of intellectual assets, as supported by financial resources, and also the continual extraction of the latent value in these intellectual assets through the effective execution of appropriate business strategies.

Figure 10.1: The Value-Creation/Value-Extraction Cycle.

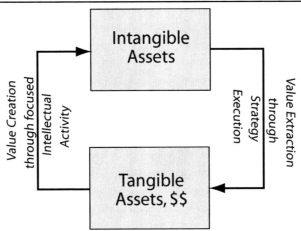

It may seem reasonable in some circumstances to say, in effect, "show me the money"—that is, to wait until the cycle is complete before placing a firm value on the result. In this view, intangible assets are only *potential* assets until they are successfully converted through successful business activities into real (tangible) assets, either physical or financial. Notwithstanding the merits of this very conservative view, there are many instances where an estimate of company value is needed *now*; waiting for alleged certainty—while being thereby certainly always wrong (since major reservoirs of value are being ignored)—is not rising to the valuation occasion.

10.2 WHAT IS THE VALUE OF A CORPORATION? —A MORE ACCURATE ANSWER

There are several approaches to the ascertainment of company value. One of these has just been discussed: the *book value*, an estimate that, if not the absolute truth, would seem to have the imprimatur of the one profession (accountancy) designated as the arbiter of such values.

For many companies, this approach is reasonable enough; for others—specifically those whose business relies primarily on the intellectual energy and output of its *knowledge-based* employees—this old-fashioned mind-set will reveal only part of the story, and not the most important part. For the sort of companies of interest to many of this book's readers, intangible assets should not be *subtracted* from the shareholders' equity (as formulated by classical accounting), they should be *included* and, indeed, they should actually be *amplified* to represent a more realistic measure of value.

What Does Mr Market Say?

When companies are publicly traded, their value is constantly under assessment and review by many potential investors. The opinions of this group—collectively known[7] as Mr Market—trump all professional opinions (including the opinions of accountants). If the company is large enough to attract professional[8] assessment, there will be stock analysts holding forth, for a fee, on the prospects[9] of these companies. While the tools and devices used by these analysts are not, for understandable reasons, exposed by their practitioners to examination by outsiders, one may assume that they include an intelligent examination of the *intellectual capital*.

The most realistic valuation of an important selection of companies, as indicated by Mr (or Mrs) Market's being willing to slam down his (or her) cash on the table to purchase shares,

---

[7]The highly effective pedagogical artifice of combining all investors as Mr Market was invented by Benjamin Graham (see his *The Intelligent Investor*, Fourth Revised Edition, HarperBusiness, 1984) and made more dramatically prominent by Graham's most celebrated (and successful) acolyte, Warren Buffet.
[8]One of the lessons of the value boom-bust phenomenon *circa* 1990 is that professional opinions of the value and prospects of companies can be highly unreliable. Whether this is due to professional incompetence, conflicts of interest or other imponderables, is a subject whose elucidation the author will leave to others. Here it will simply be stated that the current hands-off policy (i.e., paralysis) of the accounting profession with respect to the valuation of intangible assets always leaves the field open to opportunistic scammers and incompetents of many stripes to proffer their valuations of companies.
[9]These prospects focus on investment potential, meaning both anticipated growth in stock value and expected dividends.

is shown[10] in Fig 10.2. This depiction focuses on the relationship between the book value of these companies (as established by the putatively objective principles of the accounting profession) to the value of those same companies as actually held by those who are willing to buy and sell shares in these companies.[11]

Figure 10.2: Market-to-Book Value Ratio for the S&P 500.

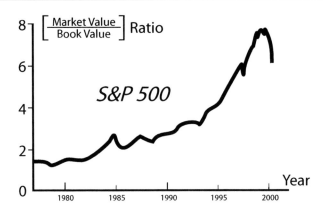

There is a conspicuous bubble forming after 1990. This is a well-recognized aberration caused partly by corporate accounting scandals and partly because investors forsook the fundamentals of sound valuation, sometimes perversely valuing website eyeballs over dollars of earnings. It is taken as admitted, as a given, and is not the point. The issue is whether the market/book ratio, over a quarter of a century, is in any way noticeably different from unity. Fig 10.2 provides a decisive answer.

Most readers of this book have a background in engineering. The author now asks this: Can you imagine any part of well-established engineering where a 100% error, or a 200%, error, or an even much greater error, would be tolerated in a bottom-line calculation? Yet, if company valuation is assumed to be one of the functions

---

[10] Prepared by Baruch Lev, Bardes Professor of Accounting and Finance, Stern School of Business, New York University.
[11] In the vernacular, "Put your money where your valuation is!"

of the accounting[12] profession, and if several hundred percent is a typical[13] error that is tolerated with nary a whimper from those who directly or indirectly pay for this sort of information, this creates a very interesting situation.

Obviously, one should search for an explanation for this colossal discrepancy. Without wishing to create either a new theory of the stock market or of company valuation (these would be side goals in the present book), it is nevertheless strongly suggested here that *intellectual capital* (to be more carefully defined in §10.3) is a major value component of many modern corporations, including most especially corporations for which readers of this book may be seeking to work, may already be working, or may be wishing to create.

Intellectual Capital

We have reached the point in the discussion where we must admit that there is concrete evidence for the claim that much of the worth of many modern companies lies, not just in their land, buildings, inventories and equipment, or in their cash, short-term investments and receivables, but in less tangible (though no less real) categories. This is especially true for companies of professional interest to engineers. We have also seen that accountancy seems not to have had any noticeable interest (or, at any rate, any success) in evaluating these intangible aspects of modern company value. Such assets are viewed by the profession as being hot potatoes—inescapably present, yet ineffectively dealt with; the result is indistinguishable from their being completely ignored.

Now we take a more detailed look at what these intangibles are, and collectively call them *intellectual capital*, for reasons made

---

[12]Perhaps the author is off-center in assuming that company valuation, meaning more than some elementary fiddling with some already-known numbers, is a reasonable output function of accounting. If so, a new profession that is able to discharge this function would be a welcome addition to the professional mix.
[13]With reference to Fig 10.1: If the *average* error, over hundreds of companies, is, say, 200% (at a particular time), we may safely assume that the error in assessing the value of some of the *individual* companies will be many times larger.

plain by Fig 10.3. In Fig 10.3 (a), we see the basic structure of the classical Balance Sheet (see earlier examples in Figs. 8.10 and 8.13): assets (A) are larger than liabilities (D) and the difference is shareholders' equity (E). This leads to an (accounting) book value that is essentially[14] E.

Figure 10.3: The Concept of Intellectual Capital.

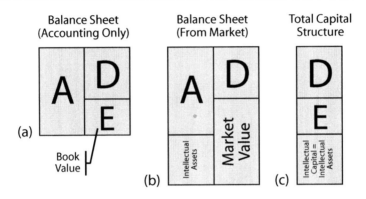

However, as shown in Fig 10.3 (b), we take note of the fact that the price people are willing to pay for the company (i.e., the market value) is larger than E—in fact, often considerably larger, as shown in Fig 10.2. If we replace the (essentially incorrect) book value by the correct (market) value, and if we insist that the Balance Sheet remain balanced, we are led inescapably to the conclusion that there must be some assets missing. There are several approaches to explaining (and labeling) these intangible missing assets, but the one of greatest interest here is that these are the *intellectual assets* we have been discussing above.

Finally, in Fig 10.3 (c), we relabel the right-side of the Balance Sheet in Fig 10.3 (b). Specifically, we break market value into two parts: book value (based on classical accounting, basically equity) plus *intellectual capital* (based on more modern views of broader

---

[14] We will ignore for this illustration that the actual accounting practice is that "book value" should have any references to intangible assets subtracted from the shareholders' equity!

scope of what creates value within corporations), which is equal in value to the intellectual assets[15] we inferred in Fig 10.3 (b).

Other Aspects Of Value

Before moving on to a more detailed discussion of the sources of intellectual capital, we should pause to remark on two more aspects of how company value might be determined. The first of these is that, while the discussion above has been based on the Balance Sheet, there are equally important indicators of value from the Income Statement and the Cash Flow Statement (see Chapter 8).

For example, it is arguable that when someone buys stock in a company (or even the whole company), they are purchasing the future earnings stream, discounted back to the instant of purchase:

$$V = \sum_{n=1}^{\infty} \frac{I_n}{(1+k)^n}$$

Here, V represents value, I_n the net income at the end of Year n, and k is the discount rate, which depends on the return needed by the investor, ultimately determined by the market, based on risk. In theory, k should vary with the economic climate from year to year, so that we should replace

$$(1+k)^n \rightarrow \prod_{m=1}^{n}(1+k_m)$$

---

[15] Readers who care about the careful use of terminology may well wonder why the author is choosing to use two terms—intellectual assets and intellectual capital—to describe what is, in effect, the same thing. Certainly, it is not absolutely necessary to have or use both terms. Still, the author has chosen to present both phraseologies, for the following reasons. First, both terms are used nowadays, so the reader should be familiar with both. Second, we tend to use intellectual assets when thinking about the left side of the Balance Sheet (since the left side is a list of assets) and we tend to use intellectual capital when thinking about the right side of the Balance Sheet (since the right side is a list of equities, or claims, on who owns the assets on the left side). Although the analogy is not perfect, it is the shareholders who also own the intellectual capital, and that is why they are willing to pay for it. Third, the word *capital* is more like a group noun (like "a load of fish") that does not cry out for enumeration, while *assets* (more like "a few fishes") seems to indicate a countable number of well defined items. Finally, *assets* is hard to dislodge from the accounting environment, while *capital* is more connected with the world of finance, which has become mathematically sophisticated and has shown itself willing and able to model anything that matters. At all events, we shall use *intellectual capital* almost exclusively from this point on.

Chapter 10: Intellectual Capital

but since nobody except possibly Nostradamus knows the exact economic climate in all future years, this flourish of accuracy is an empty promise. Come to think of it, nobody can predict even the *average* return (i.e., k) needed over all future years either, nor can anyone predict the earnings I_n in all future years. Thus, while this theory is elegant enough, it is like many results in mathematics: it explains how one can calculate something one wishes to know in terms of a number of other things one also doesn't know.

The above value formula remains true whether the earnings I_n are all paid out every year (as in investment trusts) or all used for growing the company (as in smaller, younger companies) or something in between (as in large, mature companies).

If it is a reasonable assumption that the company is at or near a steady-state performance where all the I_n are identical (= I), then

$$V = I/k$$

which, when rewritten slightly, states that the price/earnings ratio (V / I) should be V / I = 1 / k. (For example, a price/earnings of 20 corresponds to a return, or discount rate, of 5% under these assumptions.)

Another celebrated (but rather unrealistic) case where the summation can be reduced to a simple result is where the company's earnings keep growing indefinitely at a fixed rate, e.g., $I_n = (1+g)^n I_0$, where g is the growth rate and I_0 is the current trailing (last twelve months') earnings. The result is

$$V = \left[\frac{1+g}{k-g}\right] I_0$$

As should be expected, this formula blows up completely if the growth rate exceeds the discount rate ($g \geq k$).

If one sells all or part of the company just before the end of Year N then one should use

$$V = \sum_{n=1}^{N} \frac{I_n}{(1+k)^n}$$

with I_N in the sum replaced by the proceeds of the sale ($I_N \to V_N$), and, for consistency, the sale price should be

$$V_N = \sum_{m=0}^{\infty} \frac{I_{N+m}}{(1+k)^m}$$

In a similar fashion, it is arguable that when someone buys stock in a company (or the whole company), they are purchasing the entire future free cash stream, discounted back to the instant of purchase:

$$V = \sum_{n=1}^{\infty} \frac{C_n}{(1+k)^n}$$

In fact, over the entire future period, this sum and its earnings-based cousin should give the same result under consistent assumptions. This is because all the accounting can be thought of as indicating the levels of *stocks*[16] or measuring the rate of *flows*, and all the value must eventually be one place or the other.

Whether one wishes to use the present value of future earnings or the present value of future free cashflow, the question at issue here is the same: Where is intellectual capital in all of this? The answer is that it does not appear—explicitly—although one can argue that the earnings (or free cashflow) will be higher, and the risk (and therefore k) lower in a company where, other things being equal, there is more intellectual capital. Both raise the value. One can *seem* to avoid needing to consider intellectual capital with the nifty infinite sums just presented, but of course this is an illusion: one cannot actually use

---

[16]In this sentence, the word "stock" is unrelated to company ownership but is the usage of economists, who refer to *stocks* and *flows*. The analogy in electrical circuits would be to stored electrical charges in capacitors (stocks) and currents (flows). A more down-to-earth analogy would be the plumber's view: all the water must be in the tanks (stocks) or flowing through the pipes (flows); if we keep track all the storage level changes in the tanks (analogous to earnings) or all the net flows in the pipes (analogous to cashflow), we must ultimately get the same answer.

these sums without estimations of k and all the I_n (or C_n), and these will depend on all the value drivers the company possesses, foremost among which are the many components of intellectual capital.

10.3 SOURCES AND COMPONENTS OF INTELLECTUAL CAPITAL

Presumably by now the reader will have been convinced that there really is such a thing as intellectual capital,[17] either from one of the arguments above, or (as seems more likely) from his or her own observations and experience.

It would be most gratifying to report that a set of Λ well-known state variables, $\{IC_1, IC_2, \ldots, IC_\Lambda\}$, have been agreed to by the business-theoretic community as representing the intellectual capital components of a company. Even better: the total intellectual capital of any organization would be calculated through a Nobel-Prize-winning formula[18] of the sort

$$IC = \varphi\{IC_1, IC_2, \ldots, IC_\Lambda\}$$

This theory would include methods for measuring these state variables, and how they could be increased and controlled so as to lead to financial benefit. (Recall Fig 10.1.) It would be gratifying to report all these developments, but the day that this report can be made appears to be, unfortunately, a long way off.

Operating System For A Network Of Brains

Thus far, the theory used in practice by the accounting profession corresponds to the following nihilistic formulation,

$$IC_\lambda = 0, \quad \lambda = 1, \ldots, \Lambda$$

---

[17] Intellectual capital is a subset of *human* capital. The distinctions should be self-evident. (Hint: The former is primarily brains, while the latter also includes brawn.)

[18] Compare with Footnote 16 in Chapter 3. By comparison, a fully developed intellectual capital theory should get a *decade* of Nobel Prizes.

This can surely be improved upon a great deal. We shall rejoin this issue presently in §10.4.

It is helpful to recognize the simple categorization shown in the accompanying box. Surely it does not take a grand theory to discover that attracting, hiring and retaining outstanding individuals is crucial to the campaign to have the best brains working for the organization. Leading companies are keenly aware of intellectual capital concepts when hiring their key people. Even relatively junior people—from whom much is expected and for whom an exciting career progression is anticipated—are given an extraordinarily robust gamut to pass through, with several interview stages, a complex exchange of information, and a great deal of thought and effort on the part of both the hirers and the hired.

> **The Two Most Basic Categories of Intellectual Capital**
> 1. The sum of the intellectual capital residing in individuals
> 2. The many interactive (i.e., team) components to intellectual capital

In terms of the computer software metaphor, each individual mind is a macro that performs certain functions; then the operating system corresponds to all the processes and communication systems between and among these minds as they function smoothly as a team. This generally describes much that is intellectual capital.

Best Brains

What does best brains actually mean? High IQ? Certainly it would be perverse to claim that IQ[19] doesn't matter. Other things being

---

[19]Actually, as a brushing acquaintance with psychometrics reveals, the most well-known general IQ tests represent compendia of more specialized IQ measurements, including, for example, *verbal* IQ (the ability to know and understand the meanings of words); *abstract* IQ (the ability to detect relationships between patterns); *mechanical* IQ (the ability to detect, in a Rube Goldbergesque contraption of belts, gears and pulleys, what pushing on X does to Y); *three-dimensional* (or *spatial*) IQ (the ability to visualize three-dimensional objects from various two-dimensional, or planar, projections); *quantitative* IQ (that gift most-prized by mathematicians and applied scientists); and many other less well known mental abilities.

equal, the higher the IQ, the better. But there are a great many other mental talents, and the higher one ascends in an organization, the more of these other talents there are that matter. In knowledge-based businesses, the smarter someone is, the better—the company idea is to leverage the most neurons per dollar—but smart means a great deal more than it meant in high school and university.

Everyone knows people who seemed average by scholastic standards, but yet who later made scintillating contributions to their employer's organizational objectives. The academic community tends to call these individuals late bloomers, meaning that the academic standard of measurement was complete and intact, but that the vital data to reveal the true quality of these former students were not yet available *en academe*. Surely an equally likely hypothesis is that these persons simply had a set of intellectual skills that were not a perfect match for the academic demands of exams, tests and lab reports.

Sadly, there are also those who populate the opposite end of this expectation-fulfillment spectrum: they impress everyone (especially themselves) in engineering school yet they seem to be terminally destined to endure tough sledding throughout their four-plus decades in the professional workforce.

To illustrate that it is not just high levels of skill that are wanted, but skills in certain desired areas[20] and of the right type, we have Fig 10.4. Shown are two columns from a table prepared[21] by the Conference Board of Canada for the Bank of Montreal. On the left are the skills and competencies the Bank has decided to foster and encourage; on the right are the inferred *employability skills* for prospective employees. There has to be a match. Note also how many

---

[20] An interesting experiment would be to form a small company and hire a world-ranked violinist, a grandmaster chess player, a best-selling novelist, a professional football quarterback and assorted other individuals of unassailable quality, all having nothing in common either with each other or with the tasks needed by the new company, whose mission is to manufacture artificial lobsters for the mass restaurant market. Chaos and acrimony can safely be predicted. One cannot judge the skills of employees in isolation; one can only do so after convoluting their skill profiles with the profiles needed by their organization.

[21] Kitagawa K, "Bank of Montreal's 'Learning for Success'," Case Study 24, Conference Board of Canada, 1999.

of the skills needed are the so-called soft skills, and how some post-secondary curricula do not teach them very well.

Figure 10.4: Skills Needed—The Desired Outputs from the Best Brains.

BMO Capabilities and Competencies* (examples only to illustrate parallels)	CBoC Employability Skills
Skill and Knowledge Capabilities **Manage Individual Relationships** *Communicate Effectively* • Verbal and non-verbal skills • Listen, etc. *Determine Solutions* • Conduct initial assessment... • Analyze... • Develop options...etc. **Self Management** *Manage Personal Career Development* • ...engage in performance improvement... • Maximize personal contribution...etc. *Manage Change* • ...adapt... **Teamwork** *Working With Others* • ...cultivate good relations... • ...show sensitivity to co-workers' values...etc. **Behavioural Effectiveness Competencies** • Continuous Learning/Improvement • Courage and Self-Confidence • Flexibile/Resilient/Resourceful • Initiative • Integrity	**Academic Skills** *Communicate* • Understand and speak... • Listen, etc. *Think* • Think critically and act logically... • Understand and solve problems... • Use technology...etc. *Learn* • Continue to learn for life **Personal Management Skills** *Positive Attitudes and Behaviours* • Self esteem and confidence • Honesty, integrity...etc. *Responsibility* • ...set goals... • ...plan and manage time...etc. *Adaptability* • A positive attitude toward change • ...respect for diversity...etc. **Teamwork Skills** *Work With Others* • Understand and contribute to the organization's goals • Understand and work with the culture of the group, etc.

*Skill and knowledge competencies on the one hand and behavioural effectiveness competencies on the other are two manifestations of what the Conference Board calls employability that BMO measures.

The Skandia Model

Much more could be said about individual skills; about the importance of trying always to broaden oneself, especially if a career in management is being pursued; about lifelong learning; and about how to welcome and thrive on the opportunities presented by the need for change. The "best brains" approach embodies much of the intellectual capital of the company, but much is yet missing: we need to examine how these people interact with each other to produce, as a team, the needed company outputs. This operating system, too, has many parts, each of which represents a valuable module of intellectual capital.

Several approaches to describing this operating system have been suggested. We choose here to look at one in some detail; the Skandia[22] model shown in Fig 10.5.

Figure 10.5: The Skandia Model for Intellectual Capital.

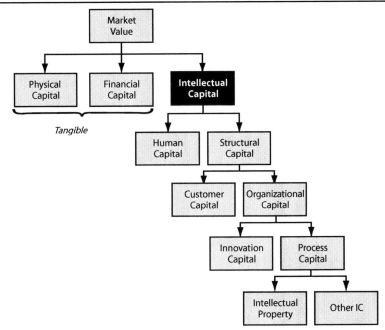

The market value of the company is shown, not just based on the *physical capital* and the *financial capital*—the two kinds of tangible assets computed from accounting—but also with the *intellectual capital* included as well. (Compare with Fig 10.3.) This, in turn, is broken down in terms of *human capital* (what we have been calling the best brains) plus *structural* (i.e., nonhuman) *capital*. The latter, in turn, is recognized as having external and internal components, the

---

[22]L Edvinsson, M Malone, *Intellectual Capital: Realizing Your Company's True Value by Finding Its Hidden Brainpower*, HarperBusiness (1997). Skandia is a Swedish insurance company.

former called[23] *customer capital* and the latter *organizational capital*. Into the latter category fall *innovation capital* and process capital, referring respectively to **(a)** the capacity and processes in the company for renewing itself and responding to change (Chapter 9), and **(b)** all the other processes and intangibles (operating systems) that help the company function at a high level.

Intellectual Property (IP)

One part of process capital is *intellectual property*,[24] which refers to the set of legal mechanisms for recognizing, rewarding, and giving exclusive property rights, for a specific period of time, to those who innovate. These legal rights encompass (through patents), how ideas are expressed (through copyright), industrial designs, brands, trademarks, trade secrets[25] and similar intangibles. These rights prevent competitors from impinging; more precisely, they give the owner the right to sue competitors who impinge. To listen to some members of the legal profession who specialize in helping with ownership of patents, trade secrets, industrial designs, copyrights and trademarks, intellectual property is the only type of intellectual capital that matters—clearly a grave oversight (and an indication of vocophilia).

Patents and other similar intellectual property protections are much more important in some situations than others. They are costly to create and they divert management's time away from what are, in many cases, even more important activities (such as setting up marketing channels). In other cases, patents and other intellectual property protections are crucial and must be obtained. Perhaps

---

[23]While a set of established customers is the single most important external category, the following are also valuable to the successful company: smoothly running supply chains, good relationships with competitors, high regard in the business sector, and other similarly positive externalities. We also note in passing that an accountancy practice would probably count its client list as its most important intangible asset—certainly that is what a buyer of the practice would be evaluating carefully—which probably explains why accountants decided to use the term [customer] *goodwill* as the excess of market value over book value for their own practices, and hence for everybody else also.

[24]Note that we must strenuously resist any tendency to confuse the terms intellectual *capital* and intellectual *property*. Such muddied thinking is clarified by Fig 10.5, which indicates that the latter is a subset of the former.

[25]"Trade secrets" are also referred to as "proprietary" or "company-confidential" information.

the best examples of the latter are found in biotechnology generally and, more particularly, in pharmaceuticals, especially where the following three conditions pertain: **(a)** the item being patented can be defined very precisely (perhaps a certain molecule), **(b)** it is relatively straightforward to copy the invention once its business potential becomes more generally known, and **(c)** many years will pass between the time of the invention and the time the economic benefits are reaped. Contrarily, if the definition of the invention is a bit blurry, and arguably indistinguishable from other similar inventions that have been or that may be made; if great cost is needed to reproduce the invention (high barrier to entry); and if the whole business opportunity will rise and fall in less time than it takes the patent to even be processed; then it would be difficult to justify the time, expense and distraction of a patent.

Patents are essentially a deal between an inventor and the society (or societies) in which the inventions will be exploited: If the inventor will disclose his intellectual property to the outside world for its wider use in perpetuity, society will in turn compensate the inventor by giving him absolute protection for an extended period of time, during which he can recapture his financial outlays for making the invention (bearing in mind that most inventions go nowhere), provide capital for continued innovational activity and provide him with the economic rent (profit) that he deserves for the use of his capital.

Finally, it should be mentioned that there are both *offensive* patents and *defensive* patents. The former are to support further development by the company, from the state of having only an embryonic invention to the state of having a fully successful commercial product (see Chapters 5 and 9). A defensive patent, by contrast, is intended to secure ownership so that one's competitors cannot benefit from the invention.

Other Intellectual Capital
The last box in Fig 10.5, labeled "Other intellectual capital," is more than just a catch-all box or a miscellaneous category. It includes all the

ways people in the organization interact, like a well-oiled machine, and all the technologies, standards, processes and policies used in such interactions that are not explicitly covered in the earlier categories.[26] For example, consider the company policy manual. Although this document is often the butt of derision for being followed too slavishly by unimaginative bureaucrats and managers, it also contains the "rules of the game" that have been codified. If this manual has been well thought out, is applied with intelligence and is generally accepted as wise and fair by the affected employees, it is a dossier of great value to the corporate team, rather like a football playbook.

Another example of "other" intellectual capital is the Six-Hat process described in §9.2. That process takes time to refine and to use effectively; in its fully functional form it is clearly a part of the team's intellectual capital. More generally, the way *all* meetings are conducted is part of the organizational operating system and if they are conducted well, this too is intellectual capital.

Accountants Are Taking Note

There is some movement within the accounting profession[27] to devoting a modicum of attention to addressing the issue of accounting for intellectual (as well as physical and financial) assets. The general approach, illustrated in Fig 10.6, is to extend their area of reporting responsibility to intangible assets as well as tangible assets.

---

[26]The important point here is not to pinpoint exactly which category something falls within, or to argue about minute distinctions—although obviously fine distinctions can sometimes be very important (e.g., in law). To spend one's time wisely, one must avoid arguing about unimportant minutiae, including falling into the grip of the "tyranny of categories." For example, most would define a liquid as "something that flows but that does not expand to fill its container." This means that all glass windows are arguably liquid (not solid) since, in all old buildings, the panes are, from gravity, somewhat thicker at the bottom than at the top. The glass is always (slowly) flowing. Interesting, perhaps, as a discussion at coffee break, but hardly a key issue to be resolved. As soon as a classification system begins to dominate and ceases to serve, get rid of the classification system. The gravamen here is that valuators should include every significant item of intellectual capital *somewhere*.

[27]In this and similar sections, the profession of accountancy is taken as represented by its publicly visible (i.e., *financial*) accounting members. There may well be companies with employees of an accountancy persuasion who *internally* evaluate intellectual assets, meaning that they produce outputs that help managers make better decisions with respect to the key processes depicted in Fig 10.1. However, these outputs are never available publicly (as financial reports are) so there is no way to comment on them.

Chapter 10: Intellectual Capital

Figure 10.6: Accountants Look at Intellectual Assets.

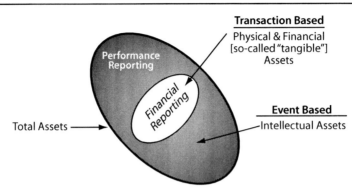

This will take some doing. The profession is exceptionally conservative, and understandably so, owing to their responsibility to providing objective (court successful) values for things. As explained earlier, this has led to a reliance on *transactions* as the most defensible measure of value in a free-market society.

From an engineer's viewpoint, on the other hand, one might ask two questions:

A. Accountants have been singularly fortunate in having virtually all their working quantities expressed in a single (financial) unit—in North America, *dollars*. Why should all their findings, throughout all their careers, be in only one unit?[28]

B. "Transactions" may be ideal, but when less-than-ideal data are available, should one just throw up one's hands and quit?

Perhaps we can anticipate a flowering of the accountancy profession in the near future.

---

[28] This would require some training, not only on the part of accountants, but also on the part of their audiences. If some ingenious accountant tried to ascertain *all* the value components of a company, including some intellectual capital entries that are not in dollars, and to report these findings to shareholders or to the Board of Directors, one can easily visualize the shrieks of frustration on the part of his or her audience: "But what is that in DOLLARS?" they would cry, missing the point entirely. (And one can only shudder at the response of banks, which do not give full credit even for *tangible* assets like inventory and accounts receivable.)

10.4 KNOWLEDGE MANAGEMENT

Classical management stressed the importance of, among other things, following the money (the Chief Accountant, the Comptroller, the Chief Financial Officer, etc.), usually to the nearest penny; attending to the physical assets (whether land, manufacturing equipment, computer and office equipment, etc.); and tending to the overall operations (executing the business plan, meaning getting the inputs in and getting the outputs out). However, now we know that for many modern companies intellectual assets can be at least as important as physical and financial assets, it must be asked: Who is looking after these intellectual assets?

Or, to put matters more dramatically: Why is it that some outfits count office inventory almost to the nearest pencil while never even glancing at the skill-sets for their employees? Why does the MIPS speed rating on computers sometimes seem more important than the productivity of the knowledge workers who use them? Why does the boss's office get a new set of furniture when key creative people are told that the company cannot afford to let them take a professional short course with obvious and immediate application to their job? These questions and innumerable other similar ones all ask the same thing: Who's minding the store on *intellectual* capital?

The Refinement Of Knowledge

Somebody should be. This newly recognized need has spawned a new type of management[29] category—*knowledge management*. Unless we return to the dark ages, this cannot be just the latest fad. One definition is shown in the accompanying box. More informally, it is the process of identifying what information the company has or can get that could benefit others in the company and then devising ways of making it available.

---

[29]The work of David Skyrme (website: www.skyrme.com), which is concerned with all aspects of this subject, including helpful techniques for the actual measurement of intellectual capital components and practical policies for knowledge management, seems right on the mark to the author.

Chapter 10: Intellectual Capital

> **Knowledge Management [Definition]:**
> The processes of assisting the creation, organization, diffusion, utilization and exploitation of *knowledge* that is *vital* to the organization.

It should not need stressing that the knowledge referred to is not limited to scientific, engineering or technical knowledge, but encompasses knowledge of all sorts relating to creating value for customers and thereby for the company. For example, one subset of business knowledge is business intelligence, which attempts to acquire (by legal means) knowledge about one's competitors (and similar strategic information).

Figure 10.7: The Stages in Refining Knowledge.

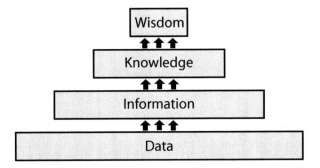

At the risk of annoying readers one more time with the definition of an important word, the significance of *knowledge* is implied by Fig 10.7. First, there is the almost infinite quantity of available *data*. (Think of an endless string of 1's and 0's.) The next challenge is to extract *information* from this huge pile of data. (Think of all the media sources, of all kinds, currently available in the western world.) Next, one must make sense of all this information—that is, information must be transformed[30] into actionable knowledge,

---

[30] Experiment: Give a top-drawer CEO yesterday's newspaper, but tell her that it is *today's* paper. Her reading will continue in an animated and enthusiastic way until someone tells her that she is, in fact, reading *yesterday's* paper. The least-violent action is that she throws the paper into the trash-bin. How can only twenty-four hours make such a difference? This illustrates the difference between information and actionable knowledge. The latter has a very short-lived "best before" date.

but this is generally impossible (except for the best abstract[31] thinkers) unless it connects with some concrete situation. In other words, knowledge is usually contextual.

The final step in the knowledge process is the realization that knowledge can eventually become *wisdom*. This last step involves not only applying information to a specific application, but in internalizing all such successful applications, in discerning patterns in the success or failure of these knowledge applications, and in developing the ability (which should be highly prized for both its importance and its rarity) to—as has been said of the greatest chess masters—flip a pawn in the air and have it land on the best square. If only it were that simple; the highest human functions are ineffable.

The trick is "to know what you need to know, but don't." Ultimately,[32] only *humans* (or their computer codes) can convert *data* into *information*; only humans can convert information into (actionable) *knowledge*; and only humans can glean the kernels from their *knowledge* to become *wise*. Only *humans*[33] have the unique ability to transform *insight* into *action*.

The Business As Control System

In the end, Fig 10.1 cannot be overstressed; it is the logo for this entire chapter—and, much more important, for many sectors of modern business. It is not enough just to collect intellectual capital; it must be properly managed and directed to the execution of an effective business strategy so as to recover financial value from that intellectual capital. A more detailed—and even more important—conceptual diagram will now be presented.

---

[31]Several organizations and books provide summaries of *best practices* for sundry business situations. These are very useful, and in many respects come as close as possible to a theory of business. However, in their raw form, these are still only *information*, because no one can know every important aspect of a particular business situation. In order to become *knowledge*, this information must be contextual, a process that requires an intelligent human brain.
[32]Some who work in the field of artificial intelligence may beg to differ with these statements. This is not at all important. We are speaking here about now and the next ten or so quarters, not about some science-fictional epoch in the distant future.
[33]And only a small fraction of humans, apparently.

Chapter 10: Intellectual Capital

If Fig 10.1 looks something like a control system, that is no accident. The notion of a control system is not an analogy; it is exactly what we are talking about. Very few readers with an engineering background will be unfamiliar with the most basic control system shown in Fig 10.8. The dashed line separates the controlled system from the outside world. One controls the system (shown in bold) by asking the following questions: What *system outputs* are available and measurable? What sensors can we use to measure them? What influences (forces) can we bring to bear on the system to influence it? How can we create those influences? And, finally, given the measurements available and the influences wanted, what would be the best control policy to convert the measurements we have to become the influences we need?

Figure 10.8: Basic Control System.

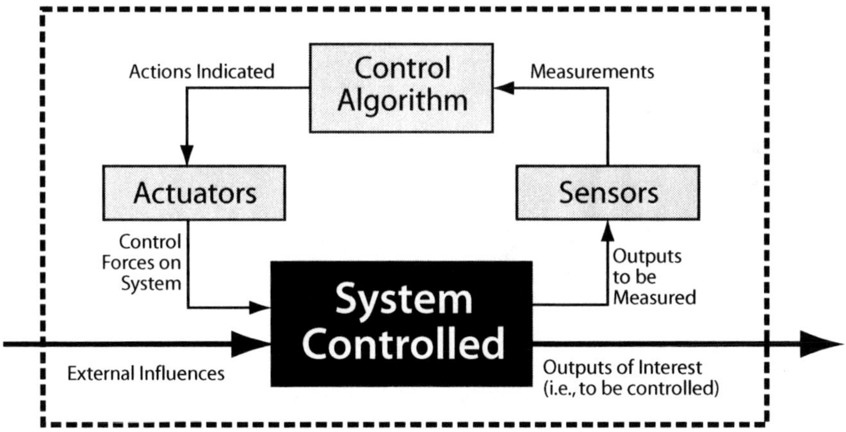

```
                    ┌──────────┐
   Actions Indicated│ Control  │ Measurements
        ┌──────────→│ Algorithm│←──────────┐
        │           └──────────┘           │
   ┌────┴────┐                       ┌─────┴────┐
   │Actuators│                       │ Sensors  │
   └────┬────┘                       └─────▲────┘
   Control│                          Outputs│
   Forces on│      ┌──────────┐      to be  │
   System   │      │  System  │      Measured
            └─────→│Controlled│─────────────┘
   ─────────────→  └──────────┘  ───────────────→
   External Influences           Outputs of Interest
                                 (i.e., to be controlled)
```

With only modest changes in terminology, and by casting our imagination into the business realm, we can adapt Fig 10.8 to the form shown in Fig 10.9 and thus have a solid basis (including literally tons of control system theory, if anyone wants to use it) for discussing the optimization of business performance.

Figure 10.9: Well-Managed Business.

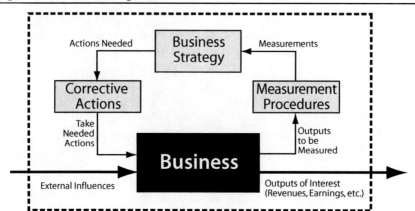

Starting with the right side of Fig 10.9 there are thousands of possible measurements that can be made, with the word "measurement" implying that the result is a *number*. Numbers, properly acquired and interpreted, are much more valuable than guesses and hand-waving. They can be added, subtracted, multiplied, divided and averaged, and they can be compared quantitatively with other companies, other years, forecasts and budgets. Traditionally, mostly financial measurements were made by managerial accountants, and these were, in addition to their internal purposes, rolled up quarterly into official financial statements for external consumption. Whatever business strategy[35] was chosen in the past, the only solid basis was largely a stream of financial data. All other management actions with respect to non-financial matters were at best intuited from experience and at worst ignored.

Since the Business in Fig.10.9 has many stocks and flows related to intellectual capital, it seems only sensible to devise *measurements* for these, to make them part of the business strategy and to influence (i.e., to attempt to change) not just financial state variables but intellectual capital state variables as well. This will lead to materi-

---

[35]We shall consider business strategy further in Chapter 13.

ally improved performance,[36] *even if only financial outputs are ultimately considered to be the important ones.* (See, once again, Fig 10.1, which is now seen as a simple but crucial diagram, representing a process going on within the block labeled Business in Fig 10.9.)

Similarly, on the left side of Fig 10.9, the outputs of the strategy are inputs to the business. They are generically called Actions, but in business the most common Actions are communications—letters, memos, conversations, meetings, etc. Other types of Action involve resource allocations (financial, personnel, other intellectual capital). Some of these Actions may be gentle, like influencing other individuals to move more in line with the company strategy; other Actions are more muscular, such as aborting a new product development or issuing a letter of dismissal.

Figure 10.9 is a powerful generic diagram of the process of constructing and executing business strategy. It assumes nothing less than a mathematical model—or, at least, an empirical model—of the entire business. Since no two businesses are identical and since, indeed, many are widely dissimilar, a sweeping and flexible approach to business modeling is required if this system thinking is to be widely adopted. Such business models[37] are extremely rare, if any complete ones exist at all, but at least we should know where we are trying ideally to go in developing more finely tuned and more optimally successful business strategies. Not just more fussing with accounting numbers, but a business model that is enhanced by the

---

[36] This point is so important that it may bear reemphasis for readers who are a bit rusty on mathematical concepts. The business in question has some financial state variables, x_f, and some intellectual capital state variables, x_i. If the only outputs that ultimately matter are financial in nature—as is typically true of a business—then system theory predicts the following plausible result: making frequent *financial* measurements for feedback control (that is, managing with the sole goal of achieving strategic business objectives that are *financial* in nature) will produce better results than the perverse approach of ignoring all financial state variables and focusing exclusively on intellectual capital state variables. However, *system theory goes on to predict that the best policy of all is to measure, and use for strategic management actions,* **both** *financial measurements* **and** *intellectual capital measurements.* This approach captures much more that is going on in the business and reveals much more about how it actually operates. It is this last management control architecture that is recommended here.

[37] The term *business model*, as more commonly used, is less demanding and more descriptive than—but very important and fully consistent with—the more system-theoretic meaning used here. The common usage simply asks: Sounds good, now how you are actually going to make money here?

many other quantities that ebb and flow within a company, including especially those associated with intellectual capital.

The strategy itself is developed by senior management and must pass the *advise and consent* process from the Board of Directors (Chapter 13).

Balanced Scorecard

It is clear from Fig 10.9 that three overall circumstances must apply:

1. At least some intellectual capital state variables—hopefully the most important ones, and the more the better—must be measured *quantitatively*.

2. The business strategy must explicitly admit these intellectual capital measurements into its decision matrix (otherwise the intellectual capital measurements are effectively ignored, and we're back to the 1980s and earlier).

3. The business strategy must indicate explicitly how this new intellectual capital information is to be used, through precise management actions, to correct and optimize the business enterprise, so as to support the strategic business goals.

These new loops are in addition to, not a substitute for, proper financial measurement and accountability.

One of the most successful methods for integrating intellectual capital into the management process is called the Balanced Scorecard,[38] first developed by Kaplan and Norton and explicated further by Niven.[39] In its classical formulation (if one can use

---

[38]The author would never have thought of calling this the Balanced Scorecard method (but then again, the author is not a trained marketeer). Perhaps those with systems training might find something like Weighted Performance Output more descriptive if less approachable.
[39]RS Kaplan RS, DP Norton, "Using the Balanced Scorecard as a Strategic Management System," *Harvard Business Review*, Jan–Feb 1996; and PR Niven, *Balanced Scorecard, Step by Step: Maximizing Performance and Maintaining Results*, John Wiley, 2002.

"classical" for something that is about a decade old) the Balanced Scorecard looks at the ideas intrinsic to Fig 10.9 (while never contemplating system theory as such) and suggests that four bundles of business outputs should be measured, as shown in the accompanying box.

> **Balanced Scorecard Measurement Bundles [Classical Version]:**
> 1. Financial
> 2. Customer
> 3. Internal Business Processes
> 4. Learning and Growth

The Financial bundle of measurements comprises all the common management accounting measurements—although managerial accountants might not use the word "measurement." These measurements (or data, or records, or schedules, or whatever anyone wishes to call them) are, at their high level, still useful as always and nothing here should be taken to mean that they should be discontinued.

The Customer bundle of measurements is an important part of intellectual capital. This has led to the word "goodwill" in accounting, regarded as some sort of surprise when a service business is sold, although of course it shouldn't be. It relates closely to customer capital in the intellectual capital breakdown of Fig 10.5. If one wishes to make quantitative measurements, one doesn't just handwave about "we value our customers," one measures on-schedule delivery, rate of returns and recalls for faults, rates or reorders from the same customers, etc.

The Internal Business Processes bundle of measurements varies in nature greatly from business to business, but is clearly a non-financial measurement class. Research and development measurements (§9.3) would certainly qualify well here—provided that, by R&D, we mean activities that are likely to lead to a successful business commercialization, not just salary deposits in researchers' bank accounts and papers in journals. There are myriad other examples as well, of which manufacturing benchmarks, production performance measures, sundry response times and divers process efficiencies are generic examples, to be applied judiciously to individual businesses as applicable and appropriate. There is almost no end to the possibilities in this category.

Finally, the Learning and Growth category. It has become a slogan for business leaders to say, "Our most important assets go home every night," or some similar sentiment, yet *employees* are still conspicuous on the Income Statements as expenses, yet invisible on their Balance Sheets as assets. They are, of course, human and intellectual assets, and should be treated as such, although the classical accounting system seemingly cannot adapt itself to even begin to compute the concept of recording these intellectual capital value components. According to the market (the ultimate arbiter of value, Fig 10.2) the total value of employees is often greater than the value of all other business assets combined.

True, human employees and their minds are not owned by the company—that would be slavery—and are thus not of interest to lawyers except as the subjects of employment contracts. Still, while ownership is an important distinction between, say, an important employee and a company car (which would most certainly be included, to the penny), this is hardly a convincing argument for ignoring human capital. Humans may not have internal combustion engines or be plugged into the electrical utility, but they have billions of neurons whose value is a greater treasure in knowledge-based businesses. How can a mere desk be meticulously accounted for, when the precious resource that sits in front of it is barely mentioned in the business model? The important point here is to be convinced that action in this area is critically necessary. Ideas for improving these measurements will surely abound once those who are sensitive to the need (and intelligent enough to respond) are convinced of the merits.

More aspects of the Balanced Scorecard method and more details on how to apply it to individual businesses, including not-for-profits and the public sector, will be left to the references.

10.5 ATTRACTING AND RETAINING QUALITY PERSONNEL

How do companies—and, more generally, organizations—get and keep good people? Unlike a building, which stands as an asset even if empty, or a financial security, which retains its value independently of who (if anybody) is employed, intellectual capital derives entirely from the humans employed in the enterprise, either through their knowledge outputs as individuals or through their operation as a team. It accordingly behooves management to ensure that the best people possible are on the company team.

This requires, first, a very careful selection process. Long gone are the days where a manager can glance at a few resumes and make an instant choice. The time and energy spent on ensuring that the best candidates are applying, that a strong interview process is developed and executed, and that a wise decision is made, are more than repaid by the saving in time and money not to have to dismiss a poor choice later. A possible severance package for the outplaced employee, the cost of the interview process for the new hire, and all the retraining of the old hire's replacement, are three examples of cost—and these don't include the general unpleasantness and the sometimes negative reactions[40] from other employees. Similarly, retaining top performers is even more important than not hiring poor performers, because in addition to all the costs of replacement just mentioned, the company has not benefited by "losing" a poor performer, but instead suffers from losing a top performer who has resigned.

Compensation

One of the most obvious subjects in the attraction and retention area is clearly the topic of compensation—although this should not be the opening salvo in a job interview! For highly placed managers, the structure of their compensation can become quite complex (see accompanying box).

---

[40]The reaction of other employees to the dismissal of an underperforming employee can be either positive or negative. If they like the person personally, they may not react well (at least temporarily), but many also realize that bad performance by one team member hurts the rest of the team.

> **Primary Components of Compensation for Senior Management**
> * Salary
> * Benefits
> * Short-Term Incentives
> * Long-Term Incentives

Let's begin with *salary*, the basic financial compensation component, paid regularly. This should match up with one's job description (JD), which it is very important to get straight at the time of hiring and to update at any subsequent times of promotion. One's salary is one's *quid pro quo* for executing one's JD to at least a satisfactory standard. All this and much else should be spelled out in an *employment agreement*, which, if balanced, will help protect both the employer and the employee. This agreement must be struck at the time of hiring; it is usually difficult to retrofit an employee agreement later that is satisfying to both sides.

It is sometimes considered socially smooth to pretend that salary doesn't matter. "I'm just interested in doing things I enjoy doing," is one of the mantras often heard[41] from technical professionals. Might this mean that they are not tuned into organizational goals? "I just want to do cutting-edge work, which is more important than money," is another oft-heard remark. Does this reveal a plan to stay for a few years, to learn as much as possible, and then leave?

The wise manager realizes that salary does matter. True, other job attributes may also be important, but the company salary structure should be competitive if one wishes to avoid a high turnover rate among one's top performers. One must also avoid the tendency to create a comfortable refuge for the bottom performers. What one should wish, surely, is the reverse: for the top performers to

---

[41] The author has several times heard engineers, professors or managers say, at their retirement dinners, "I enjoyed every minute of it and I'd have done it all for nothing." This is the ultimate statement that salary doesn't matter. Strangely, however, none of these individuals backed up this claimed sentiment by presenting their former organizations with a personal check in reimbursement for the then-value of the total of all the salary previously paid to them.

stay and the bottom performers to leave. This will not happen with a feeble salary policy.

Benefits

There is a vast *smorgasbord* of possible benefits. (Some of these benefits will be required by statute.) Packages usually include life insurance, disability insurance, health insurance and government pension (social insurance or social security). Larger firms can afford to get into the highly complex business of offering a company pension plan. All these benefits, and numerous other more specialized ones, can quickly amount to 10%–20% of an individual's salary, or even more.

Vacation policy is another area where employee needs and employer requirements must be balanced through careful thought. Unlike assembly line workers, whose output is conspicuously visible and easily measurable, knowledge workers can spend many hours (months, actually) very much in their offices, and very much not on vacation, yet contributing nothing at all of value to the organization that pays them. It is hard to imagine a fact of life more important to the management of knowledge workers, yet it is often sloughed off and ignored. This will be discussed more fully in the next chapter, but the point here is that whether a knowledge worker's vacation is four weeks or five is not nearly as important as the number and quality of her contributions during the 47 or 48 weeks that she is *not* on vacation.

Assistance with low-interest home mortgages or with a company car can be the linchpin in attracting key personnel, especially if they are moving from low real-estate values to higher ones, or if they have a young family to consider. Management can become quite creative if it chooses to fine-tune these unusual fringe benefits. Those who are negotiating a new position should also take the time to acquaint themselves with these and other possibilities. Permitting flexi-hours, or job sharing, or both, can also be nearly cost-free benefits that will be welcome policies for some types of employees.

Short-term Incentives

Senior managers who are in a position to have a large beneficial effect on company growth, market share, earnings, etc., through their personal hard work and sharp decisions are often provided with a *short-term incentive* (STI) component[42] to their compensation package. The intention is that managers go well beyond the merely satisfactory performance of their duties (as specified in their job descriptions and for which they are already compensated by their salaries) and reach still further.

The idea is shown in more detail in Fig 10.10. The MAX objectives, or goals, should be *stretch* goals, meaning that they are difficult to reach, and in some years may not be reached, but they should be within the realm of realistic possibility. Some goals may be financial for the current year, while others (like getting a foreign subsidiary into operation) are non-financial and have longer time horizons. One begins by setting aside an overall amount, say $50,000, as the maximum total STI for the current year. Each STI component is given a weighting factor, according to its importance, with the sum of all these fractions adding to 1.0. If, for the n-th goal, the MIN value is not achieved, the STI reward for that component is zero; if the MAX value is achieved, the STI reward for the n-th component is w_n times the total STI amount set aside ($50,000$w_n$ in our example). For achieving goal values between MIN and MAX, simple (linear) interpolation is used. If the MAX value for every goal is attained, the maximum short-term incentive (e.g., $50,000) is earned.[43]

Setting the goals and deciding on the weights and the MIN and MAX values—in other words, populating the STI grid as in Fig 10.10—is a delicate balancing act. In reality, the manager in question knows that if there were no STI system, she would probably have

---

[42]This STI package is often colloquially called the "bonus" but we prefer to eschew that appellation here because it can quickly devalue the true meaning of the STI. An Xmas turkey is a bonus; STI's are, if designed properly, well earned by hard work.
[43]If goals greater than MAX are reached, this is a pleasant problem for the compensation committee.

a higher salary. This means that she is, in effect, partially funding the STI system with some foregone salary. This is a conscious agreement between her and the company: she will give up a portion of her salary, with its somewhat success-independent, *entitlement* tone, on the basis that, with hard work, she can, on average, recover the forgone salary *and more*. Thus, if the goals are so challenging that they are scarcely ever reached—indeed, if, on average, the manager does not achieve even the level of salary foregone (despite higher risk and harder work)—the system will collapse and be seen as just a ploy to pay a lower salary.

Figure 10.10: Typical Structure of a Short-Term-Incentive Compensation Grid.

STI Grid for I. M. Manager

	Objective	Weight	MIN	MAX
Goal 1	Company Revenue	W_1	$156M	$167M
Goal 2	Complete South American Subsidiary	W_2	by Oct	by Jul
⋮	⋮	⋮	⋮	⋮
Goal N	Company Earnings	W_N	$12M	$18M

At the other extreme, if the goals and MIN and MAX levels are too soft, the whole STI system will be become just a way of shoveling more money towards the top managers, in effect paying them twice for the same work. The STI system is outstanding in theory but requires fine-tuning and commitment by both the payers and the payees in order to be successful in practice.

Stock Options

The *long*-term incentive component—usually stock or stock options—is also complex to implement, but very valuable when it works well. A *stock option* award has the parameters shown in the accompanying box.

> **Key Parameters of a (Compensation) Award of Stock Options for Senior Management**
>
> * Date of Award.
> * Number of new shares involved.
> * Price at which these new shares can be exercised.
> * Earliest date at which these new shares can be exercised.
> * Date at which the options expire.

The way stock options work is that a specified number of shares—not someone else's shares but new shares issued from the company treasury—can be purchased by the compensated manager, at a specified price (called the *strike* price or the *exercise* price), on or after a specified date (the *strike* date or *exercise* date), up to but not after a much later specified date (the *expiry* date). Obviously, if the share price becomes higher than the exercise price, these options can be exercised to a financial advantage; on the other hand, if the share price is lower than the exercise price, the options are said to be "underwater" or "out of the money" and would not be exercised.

Figure 10.11 shows the general idea. To keep things graphically simple, the vesting period is just one year, although in reality that period would be longer to make this compensation component truly a *long*-term incentive. On the top diagram is shown a case where the stock price keeps going up and the value of the options awarded each year—*with their strike price specified to be the market price at the time of award*—also continues to be positive. On the bottom is a less happy case: after three years of growth, the stock takes a dive and the options awarded at the beginning of Year 4 are underwater by the end of the year.

Figure 10.11: Stock Options Acquire Higher Value as Stock Price Goes Up.

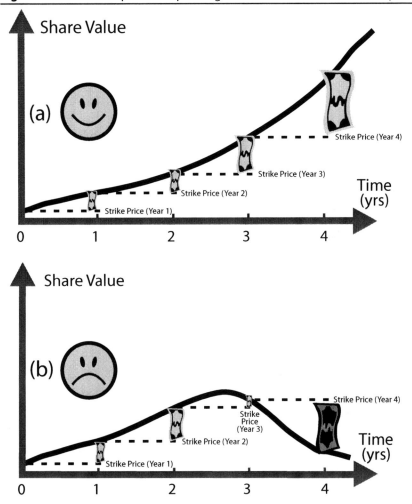

In order for options to be regarded as a successful method of compensation, it is not enough that the stock price go up. It must go up more than the amount it would have gone up without the option awards. More precisely, it must go up fast enough that the dilution effect on all the other shareholders, when these options are eventually exercised, is more than compensated for by the *extra* (i.e., more than without options) bounce in the stock.

PART III

Engineers Who Become Leaders and Senior Managers

Do You Have What It Takes?

You have now embraced the practice of management as an essential, unifying function within your organization. In fact, you consider yourself a part of management. Stated another way, you dare to say that you are, in fact, a *manager*—not in a trivial *supervisory* sense, not in a minor *administrative* sense, and not in a somewhat resentful, part-time *when I can't do "real" engineering* sense, but in a proud, *I'm making a big difference around here* sense.

You are, in other words, a senior engineering manager. The question is whether you wish always to have the "engineering" modifier as part of your job title and function, or whether you have set your sights on even higher ground.

Chapter 11 discusses the role of leadership. All leaders in modern organizations have a home profession. Engineering makes a fine such home from which to rise to much higher positions—although much more needs to be added to arrive at leadership and senior management.

One way to lead is to create the organization yourself (Chapter 12). This will be successful only if you are naturally an entrepreneur. The joy of being your own boss also has many attendant irritants. The most successful entrepreneurs get their joy from growing a worthwhile organization.

Finally, Chapter 13 examines the highest positions of management and leadership within a business organization, namely, senior management and governance. With the ubiquity and importance of high technology in modern companies, engineers who are also seasoned senior managers should be welcome in the boardroom.

CHAPTER 11

Leadership

If You Have to Tell People You Are, You Aren't

Author's Note: *In this chapter, I again have the pleasure of acknowledging my sometime co-author, Neeraj Ghai, who was introduced at the beginning of Chapter 6. His youthful enthusiasm and on-the-job experience are catalysts for many of the topics considered here. Neeraj wrote the draft of §11.3.*

This capstone viewpoint in this book—that engineers can, in principle, become outstanding managers—is obviously beneficial to engineers of all ages and stripes and especially salutary to any engineer's consideration of his or her future career options. For all this importance, this viewpoint is difficult to find in the prevailing professional literature and culture. Most engineering schools just don't promote it (Chapter 2). Few engineering societies or associations foster a direct interest in this career direction; perhaps understandably, these societies feel responsible primarily for the shepherding of engineers *qua* engineers. (See Fig 11.1, later.) And the corporate cultures of the thousands of high-performance organizations that rely on many key individuals whose core profession is engineering do not generally assume that these people are suitable for (or, indeed, interested in) more senior leadership roles.

In this chapter, we shall explore more directly the vision of individuals who start their professional lives as engineers yet who metamorphose into true managers. This evolution of performance will require more than an engineering degree and engineering experience, of course. It will take inclination, hard work, further study, a keen sense of observation and many other skills and aptitudes. In this latter connection, it will require virtuosity in the soft skills. Managers tend to spend most of their time dealing with people and people issues. They no longer puzzle over designing physical products, developing manufacturing processes, analyzing systems, solving equations or making technical innovations.

11.1 SOME ENGINEERS MAKE OUTSTANDING MANAGERS

It would be foolish to suggest that *all* engineers should plunge headlong into a quest for managerial celebrity—foolish, indeed, for two reasons. First, the engineering profession itself is highly important to society; from both an organizational and a societal perspective, having all engineers or all managers is silly. Second, anyone who has spent a career observing engineers being great engineers and managers being great managers knows that the idea of all great engineers being great management material is absurd on its face, as is the idea that all great managers (with an original engineering degree) might have been great engineers.

A Helpful Analogy

To further distinguish between the two points being made, which sound superficially and confusingly alike, consider the following two analogies:

> First, picture a football team consisting entirely of defensive centers and a second football team consisting entirely of quarterbacks. In these two strange teams, assume that *all the quarterbacks are star quarterbacks* and that *all the defensive centers are star defensive centers*.

Chapter 11: Leadership

Yet neither team could beat a good solid football squad. The reason should be obvious: It takes an appropriate division of labor and a finely tuned organizational structure to build a winning team. The analogy to business (or to society at large, for that matter) should be clear.

The second analogy is somewhat similar but fundamentally different—and closer to a realistic concern for individual players. Picture a football team that has a set of players, one for each position, with each player being quite proficient at his position. A few may even be stars at their positions. One season, there is a player on the team, a defensive center, who believes he is quarterback material. He insists on trying out for that position and he ends up successfully becoming backup quarterback. In three years, the first-string quarterback retires and our former defensive-end becomes starting quarterback. This is a success story (unless he was a great defensive end and an indifferent quarterback). If, however, all the successful, well-positioned players on the team become weary of their lot and decide that they are interested only in the quarterback position, this would destroy both the team and most of their own careers.

Similarly, not all engineers should aspire to top management; however, more should do so than do now. If this trend is to gain traction, some must start young and work diligently towards this objective.

Thus, while some engineers—more, it seems likely, than at present—may well be cut out for management, many more engineers would be better suited to perfecting their roles as engineers (although even these roles inevitably contain a management component). The point of this book is to point *some* engineers toward a possible career in senior management—whether at the student level or later on in their careers—especially those who might not otherwise seriously consider and work toward, yet who would excel at, this career choice.

Anti-management Bias In The Engineering-school Culture

Chapter 2 was devoted to the phenomenon in which many engineering schools have become, over the past many decades, somewhat snooty with respect to training the next generation of engineers. According to this ideology, the loftiest position to which a young entrant to an engineering school can aspire—after almost a decade (and a few degrees) in university—is that of *applied scientist*. With a doctorate in engineering (i.e., in applied science) the only career of real value (so the story goes) can now begin. The preferable subspecies of applied scientist is (surprise!) the university professor of engineering, who is judged largely by science paradigms—i.e., number of papers (liberally peppered with mathematical flourishes) in obscure journals.

Second best, from the engineering-school cultural perspective, is an applied scientist who graduates (also with a PhD) but who makes his or her home in a non-academic setting. Third best—but, paradoxically, by far the largest cohort of graduates—are engineering students who find positions as (uh) engineers.

Thus, early on in one's first encounter with the engineering profession, one is somewhat torn between becoming an applied scientist (although this is not the terminology one normally hears) and a mere engineer. While this is the primary predicament regarding professional identity swirling around the next generation of engineers, the *raison d'être* of Chapter 2 was to shine a spotlight on what should be, in the long term, an equally important issue for engineering students: Should a career in management be given serious consideration—not meaning supervising a junior engineer by age 40, but meaning becoming CEO of a major technology-based corporation by age 45? This career option is difficult to discern in most engineering schools.

Playing The Management Card In Mid-career

The contrived strangeness with respect to the scenario of engineers seriously entering management pervades all aspects of engineering *en academe*. Even the curricula of so-called professional development courses in most engineering schools, created as part of life-long learning, are all about *being a better engineer*—while continuing to report to other individuals in higher levels of management within the course-taker's organization (who are likely from some other profession). Not infrequently, the occupants of these higher management levels, whose dictates the engineers taking the professional development courses are striving to serve, were individuals that the engineers in question passed like a rocket back in senior high school. What happened? How did these talented engineers get passed on the corporate ladder? Perhaps they didn't know (or didn't care) that they were *on* a ladder? Anyone who reads this book will by now be familiar with these issues.

Chapter 1 (more specifically, §1.2) showed clearly that real engineers in their actual jobs perform many functions that should be classed as management, and that the older and the more capable these engineers are, the more their daily work is occupied by what only can, to be fair, be called management. Most engineers, as they gain experience and develop additional skills in the workplace, engage increasingly in management functions. Management, whether consciously chosen as a career focus or not—indeed, even if it is resented and unwelcome—becomes increasingly and ineluctably part of an engineer's palate of activities.

Some engineers may reason thus: If I am going to be spending so much time on management, I may as well face this fact head on and become a successful senior manager rather than be a part-time engineer who is a part-time manager. Other engineers may decide that they would rather focus on attaining more senior management jobs themselves than interminably whining about the shortcomings and weaknesses of the managers to whom they report. More than one person has decided that they would rather *be* the boss than be jerked around *by* the boss.[1] This is especially true for engineers, who, through their training and skills, have some *bona fides* to become "the boss."

---

[1] The Bard of Avon caused Hamlet to ask the following immortal question:" ...whether 'tis nobler in the mind to suffer the slings and arrows of outrageous fortune or to take arms against a sea of troubles, and, by opposing, end them?" The career translation is this:" ...whether 'tis preferable to stay in one's present position reporting to one's present boss (or to one's next likely boss) or to set one's career focus on the next management level up and thus be the boss?"

Engineers Becoming Managers: From the Classroom to the Boardroom

Anti-management Bias In Typical Engineering Associations

One way to examine the ascending levels of management responsibility in an engineering context is to examine the classification system used by the major licensing engineering societies that govern the practice of professional engineering. The author is most familiar with the Professional Engineers of Ontario (PEO), his own association for the past four-plus decades, and this will serve as proxy for other similar jurisdictions.

PEO has identified six Levels of Responsibility, designated Levels A through F (see Fig 11.1).[2] Note, however, that since this system of position gradations is created primarily to suggest fair salary levels, and is thus based ultimately on the level of value creation within the engineering profession (in the narrow sense), it not intended to be about the "level of management."

However, as one examines these Responsibility Levels, the hallmarks of the transition to management are impossible to suppress, despite the PEO's formidable feats of euphemism. There is (and should be) room to improve and progress in the PEO scale, not just by replacing engineering duties with management functions, but also by becoming more valuable in whatever engineering capacity one performs. Indeed, the reader has to sympathize with the creators of the table in Fig 11.1: not only are they grappling with the largely incompatible careers of normative (narrow-sense) engineering vs. highly specialized, possibly research-oriented engineering, they also have to deal with the set of engineers whose careers are clearly oriented towards management.[3]

In the first case (engineer vs. applied scientist or engineering specialist), they handle it with the preposition "OR." In the latter

---

[2]PEO Salary Survey, 2003. The Levels of Responsibility are more colloquially referred to as Rungs on the Ladder. Actually, PEO also mentions a level titled Beyond Level F, for which even PEO's verbal virtuosity could no longer provide euphemisms for "management."

[3]In Fig 11.1, there is no management career stream *per se*. To the contrary, one can only infer with some difficulty the increasing management responsibility that engineers undertake with each higher level of responsibility in the company.

Chapter 11: Leadership

	Duties	Decisions and Commitments	Supervision Received	Leadership, Authority or Supervision Exercised
Level F	Engineering administration, helps **determine basic operating policies**, meets objectives in an economical manner, OR acts as a prestigious engineering consultant.	Makes **decisions on all matters**, including policy and expenditures **of large sums**, or implementation of **major programs**, subject only to overall company controls.	Receives administrative direction based on overall policy and objectives. Work reviewed for coordination with other functions.	Reviews technical work, responsible for **schedule and program** objectives. Recommends selection, training and remuneration of staff.
Level E	Participates in short-and long-range planning. Makes independent decisions. Originality and ingenuity required for practical and economic solutions. May supervise large groups or highly specialized small groups.	Decisions not usually reviewed technically, except for **capital budget** or **strategic** issues.	Work assigned as broad objectives. Reviewed for policy, soundness of approach and general effectiveness.	Coordinates programs. Directs use of equipment and material. Recommends selection, training and remuneration of staff.
Level D	Now direct supervision of other engineers is normal, OR is a recognized specialist. Maturity in planning and conducting projects, including **innovation** and coordination.	Recommendations reviewed, but usually accepted and conclusions. Difficult decisions referred upwards.	Work assigned in terms of objectives, relative priorities and critical impingements. Work carried out within broad guidelines.	Assigns and outlines work. Reviews work for technical accuracy. Supervision may call for recommendations regarding staff selection, training, **evaluation** and discipline.
Level C	Now a fully qualified engineer. Responsible engineering assignments. Knows cross-effects with other fields. May modify procedures somewhat. Participates in planning.	May study independently, making analysis and conclusions. Difficult decisions referred upwards.	Work not supervised in detail, although guidance available where needed.	May give technical guidance to technicians or younger engineers. Supervision of other engineers otherwise abnormal.
Level B	More advanced engineering training; assists more senior Engineers.	Decisions within established guidelines.	Results reviewed in detail.	May give technical guidance to a very few engineers or technicians on same project.
Level A	Received training; simple tasks (always checked).	Decisions technical, routine and checked.	Closely supervised.	May check work of a few technicians or "helpers."

Figure 11.1: Definitions of Seniority Levels for Engineers, from one engineering association (abstracted by the author).

case (engineer vs. engineer/manager) they try to ignore it—quite unsuccessfully, if one knows how to read the tealeaves.

But how can one explain the virtually obsessive avoidance of the word "management"? It is quite evident that at least some Level C engineers are testing the management waters; that at least some Level D engineers are beginning to spend a significant fraction of their time in junior management; that at least some Level E engineers are now managers virtually full-time; and that Level F engineers are full-fledged engineering managers. This is, of course, completely consistent with the data in §1.2.

Calling A Duck A Duck[4]

The Levels in Fig 11.1 are identified, reasonably enough, in terms of four[5] parameters, thus: **(1)** duties; **(2)** recommendations, decisions and commitments; **(3)** supervision received; and **(4)** leadership authority and/or supervision exercised. Thus, in Fig 11.1, we have six Levels, each defined in terms of that four-parameter space, for a complete matrix of 24 elements. The PEO Level definitions regard Levels A and B as essentially post-engineering school training stages, where the young engineer is undergoing intense on-the-job training. Fair enough.

It is the curious aversion to standard business terminology that is the mystery in Fig 11.1. Consider the words used by the career table (on which Fig 11.1 is based):

> [From Level C]: " . . . participates in planning . . . ," " . . . may give technical guidance to engineers of less standing or to technicians . . . "
>
> [From Level D]: " . . . the first level of direct and sustained supervision of other professional engineers . . . "

---

[4]This refers to the well-known Duck Experiment in which, if a bird looks, walks and quacks like a duck, one can reasonably infer that this particular bird is a duck.
[5]There is also a fifth parameter in the original PEO chart—entrance qualifications—but after the first level or two this information is so banal that it has not been shown in Fig 11.1. Everyone involved, apparently, should have an engineering degree (surprise). The concepts of professional development and lifelong learning are not explicitly present in the PEO chart; however, it would be impossible to catalog this chaos of possibilities.

Chapter 11: Leadership

Words like "planning" and "sustained supervision," to an unbiased ear, sound like early management functions. For eyes and ears that are attuned, the sights and sounds of ducks walking and quacking are highly suggestive.

At Levels E and F, we have a veritable cacophony of quacking ducks:

> [From Level E]: " . . . short- and long-range planning . . . , " " . . . may supervise large groups . . . ," " . . . supervision of other engineers is normal . . . ," " . . . may exercise authority over a group of highly-qualified professional personnel engaged in complex technical applications . . . ," "originality required for economic solutions . . . ," " . . . makes responsible decisions not usually subject to technical review on all matters assigned, except those involving [capital budget] or [strategic] objectives . . . " " . . . work is assigned only in terms of broad objectives to be accomplished, and is reviewed for policy, soundness of approach, and general effectiveness . . . "
>
> [From Level F]: " . . . usually responsible for an engineering administrative [sic] function, directing several [professional and other] groups engaged in inter-related engineering responsibilities . . . ," " . . . determines [basic operating] policies, devises ways of reaching program objectives in the most economical manner . . . ," " . . . makes responsible decisions on all matters, including: the establishment of policies, the expenditure of large sums of money, and/or the implementation of major programs—subject only to overall company policy and financial controls . . . ," " . . . selects, schedules and coordinates to obtain program objectives . . . and/or as an administrator [sic], makes decisions concerning selection, training, rating, discipline and remuneration of staff . . . "

Let's review the evidence (Levels E and F only):

> In Level E, the bird is skilled at walking like a duck, is talented at quacking like a duck, and looks exactly like a duck. It is responsible for both short-term and long-term planning; it supervises large and small groups in complex technical work; it must be budget conscious (unsuccessfully obscured by the reference to economics); it is totally

in charge technically (technical work not usually reviewed), although, like everyone else in the company, this bird must explain how its decisions are compatible with capital budget constraints and company strategy. (The bird, of course, must be "effective.")

By Level F, this proto-duck satisfies every avian criterion for being a duck. He no longer directs several *people*; he or she directs several *groups*. He sets policy and guides programs towards their objectives with an eye toward the budget. Level F engineers—so-called at this point only because of their first degree—implement major[6] programs, responsible only to the most senior management (strategic policy and capital budget). Their only limits include the highest constraints, those imposed by the most senior management team and the Board of Directors. Notably, these Level F birds also exercise complete control over their staffs.

This review completes the Duck Experiment. The conclusion (especially with respect to levels E and F) is inescapably this: These birds are ducks! It is almost[7] humorous that all the known euphemisms for management have been used or proxied in the above PEO descriptions—*without the word "management" ever once being mentioned!*

Avoidance (in the job descriptions offered by professional engineering societies) of the frank fact that management is an increasingly important part of an engineer's responsibilities as his or her career develops could be the subject of a sociological study, but we shall leave this inquiry to others. One might speculate that it is a cultural bias of some sort; in any case, young engineers who peruse this PEO material could be forgiven for thinking that management was not a strong career path for them.[8]

---

[6]For those readers who are just scanning and not paying detailed attention, note the language here. A program is said to have several (perhaps up to ten) subsidiary projects. The person being described as Level F is responsible for (at least) one (and possibly more) programs. He or she is responsible to senior management (defined for a large firm in Chapter 13) only for overall corporate strategy, policy and budget.
[7]Almost, were it not for the fact that the careers of many worthy engineers have been kicked to the curb by the biased language of career level descriptions like those of Fig 11.1.
[8]Why would "real" engineers want to settle for management? Just as, in engineering school, why would "real" applied scientists want to settle for engineering (Chapter 2)? Note the two cultural crevasses engineering students must navigate if they wish to pursue a management career.

Even more important, in the context of the present chapter, is that there is no hint that *senior* management is a lively option for members of the engineering profession. Perhaps such guidance is beyond the obligations of professional societies devoted to the care and feeding of a different sort of bird.

11.2 BUSINESS LEADERSHIP

In this section we shall examine the concept of leadership generically. In §11.3 we shall look in more detail at management and leadership in an organizational and business setting, highlighting especially those characteristics that will help engineers (or engineering managers) decide whether to pursue a management career more intensively.

Leadership Is More Than Management

Leadership and management, though related, are not quite the same thing. It is possibly, generally, to be a manager without being a leader. Someone who goes about her daily duties, including the supervision of others lower down in the organization, and who has some responsibility for money and schedule, can reasonably claim to be a manager. Further, this someone, if she carries out her responsibilities well, can claim to be a good manager. (If not, she may already[9] be in career danger.) Engineering Levels D, E and F in Fig 11.1 are typical of management roles for engineers (meaning, engineers who have also become engineering managers). Later on in this chapter (§11.3), we shall peel back more layers of the onion to discover the characteristics of good managers. It is also possible to be a leader without being a manager.

---

[9] We have the computer revolution to thank for exposing the managers—particularly middle managers—who were (some combination of) uncreative, unimaginative, incompetent, or not involved with their staff teams on the human level. When your relationship with your IN-OUT box can be replaced by a piece of computer software costing less than $500, it is time to suck it up and either (a) improve really fast, or (b) retire. Or wait until the ax falls. Interestingly, while many traditional *management* functions are now carried out by software, the author has yet to hear of anyone offering to sell a computer program for business called LEADERSHIP.

Whether a leader or a manager, it is possible to be a good one or a bad one. The following well-known example may suffice to illustrate:

> The Rev. (i.e., "one who is revered") Jim Jones was a leader. He had two university degrees; he formed important religious organizations; and he was loved by those in his group (cult?). He convinced persons of more-or-less normal intelligence to **(a)** give his church all their worldly goods, **(b)** move to Jonestown, Guyana, and **(c)** drink, *en masse*, a poisonous liquid that bestowed, simultaneously, suicide on all the adults and infanticide on all the children. This takes astonishing leadership skills. Few, however, would think Jones a good leader.

A more light-hearted example of good and bad leadership is exhibited in Fig 11.2.

Figure 11.2: Leadership Varies in Quality. Top: Good Leadership. Bottom: Not-Quite-So-Good Leadership.

Chapter 11: Leadership

In business leadership, *leadership includes management*, but with an additional layer that is most important.

Leadership Defined

It is reasonable to ask at this point: What is leadership? Many answers to this question have been given, and the author has read quite a few of them, but the only one that tends to stick is shown in the accompanying box.

> **Leader (Definition)**
> A leader is someone that people tend to follow.

This sounds too simple. Others may say it is a circular definition. But simple is good, and surely the definition is prescriptive, not circular, if by leader one means a *natural* leader. Someone may be *imposed* as a leader, but the definition given is not circular for a natural leader.

Readers may wish to cast their thoughts back to when they were quite young. Who was the kid who led in the neighborhood group? In the classroom? At recess? At summer camp? On the sports team? Chances are it was a young colleague who had the same age and the same experience as the reader at that age. This seemingly spontaneous leader had no special authority, no job title—in short, no credentials to be leader, except the *gift of leadership*. For such people, their colleagues simply tend to follow them.[10]

This is not all good news (unless one is a natural leader). The impression may be gained from the above remarks that leadership

---

[10] If one attends a conference roundtable, with no pre-established pecking order, no detailed agenda, no résumés in the conference program—only individuals who want to have a successful result for their time spent—a "leader" will always appear, like a ghost on Halloween, within a half-hour. *Any* time a group of adults sit around a table with no organizational chart for guidance, a leader will always emerge. It is a process fascinating to observe. Some would-be leaders will also be in evidence, sometimes stating their credentials almost plaintively, as if to say, "Come on, guys, I should really be the one to lead this." To no avail, of course. Occasionally, the wrong leader is produced by this process; more often than not, it's exactly the right person—the leader of *this* group, on *this* occasion, in *this* context!

cannot be taught and cannot be learned. To a large extent this is true. When the ancients used the word we now know as charisma they thought it a divine gift. We may now know more about genetics and early nurture than the ancients did, but the idea that one either has it or doesn't still has a large measure of truth. Many of the skills of management can be taught more easily than the skills of leadership. Still, just as there are actions one can take to improve one's management skills, there are also actions (although probably fewer of them) one can take to improve one's leadership skills. These actions will be more easily assessed when the skill set required for leadership is discussed in more detail, in §11.3.

Leadership And Management—Good And Bad

One can thus distinguish between management and leadership in their generic senses, and both can be either competent or incompetent. This leads to a natural four-box, shown in Fig 11.3.[11]

Figure 11.3: Leadership-Management Terrain: Good and Bad.

[11]B-Schools (meaning, of course, *business* schools, not schools of less-than-stellar merit) love these 2 x 2 planar charts (or *four-boxes*) showing the basic relationships between two variables. This is, to business students, what the *x–y* plot is to engineering students—the basic tool to express relationships in a world that has 1,000 variables before breakfast. The author has created the four-box shown in Fig 11.3, because it illustrates several truths quite well. For engineers, a third dimension (largely independent) is performance as an engineer (narrow definition).

If leadership and management are separate qualities—although obviously preferred in partnership—one can assess one's career using Fig 11.3.

Intellectual Capital Revisited
In the last chapter we saw that many of the characteristics of leading businesses have changed over the past generation. This can be expressed as the ascendancy of *intellectual* assets relative to *physical* and *financial* assets, but this is only one verbiage for the revolution. The employees of many of the most exciting companies have changed; they are no longer useful primarily for their muscles (the archetypical physical laborers) or for their ability to process routine decisions in terms of the policy manual (the archetypical office workers). The physical labor is all done by machines and the routine thinking is all done by that most amazing of machines, the computer.

The employees of interest to engineers are now knowledge workers. They are well educated; they are fussy about for whom they work; they expect to be accorded a place of suitable prominence in the corporate hierarchy; and they decide, either consciously or unconsciously, whether to blaze a trail or to hide in the bushes and await better opportunities. For the proverbial ditch-digger, it is easy to measure the height and depth and length of the ditch; for laborers on the production line, it can be discerned how many units have been assembled in a given time; and for denizens of the old-style office, one can measure the throughput from the inbox to the outbox. For the knowledge worker, on the other hand, it is quite difficult to measure how productive he or she has been, as compared to how productive he or she *might have been* under better management—and especially under better leadership.

To put matters another way, there have in the past few decades been many technological innovations that have been of enormous assistance to managers—so much assistance, in fact, that many of these managers are no longer required! No such development has yet occurred with respect to leadership. Thus (and the author assumes here that most readers have the technical training to appreciate the phrase) the *leadership-to-management ratio* needed in business has increased markedly in the recent past. To stress these important points once again, there are not one, but two, general reasons for this increased stress on leadership relative to simple management: **(a)** many management functions have been made more efficient by technology (specifically, information and computer technology); and **(b)** typical employees today need leadership more than they need management.

Special Problems (And Benefits) With Engineers Becoming Managers
Another word of caution is needed here with respect to engineers who are on the management career path. Although, as stated repeatedly, technology is now of vast assistance in management functions, experience shows that technology is not the place to start with respect to knowledge management (or to the management of knowledge workers). Buying a roomful of computers and software and then trying to figure out how one might use them is exactly backwards. Better to decide what the knowledge management (KM) strategy should be and then apply suitable technology to support this strategy.[12]

One area where engineers in management should have a great advantage over other source professions is with respect to the special matter of how to handle technical innovation within the business (and, more generally, knowledge workers). This is one area where a highly science-based, engineering school background should prove beneficial: Such graduates should know how R&D personnel think, and thus how to manage them—although the management of R&D personnel in particular, and knowledge workers in general, is not a straightforward matter.

---

[12]This is probably an unnatural impulse for most engineers, particularly engineers whose specialty is in information technology (IT). Thus the need for this warning, which is to guard against an instinctive response that is not consistent with our basic definition of technology (see §5.4).

Chapter 11: Leadership

Apart from the occasional genius—regarding whom the best policy is probably to help all one reasonably can and otherwise stay out of the way—it takes many years to identify those who are talented researchers, or accomplished designers, or star salespeople. Given this fact, how should a business allocate its research resources? If we allocate all our resources to the research stars, on what basis can we expect the next generation of researchers to develop? (Indeed, how will we be able even to detect the stars of the next generation?) If, on the other hand, we award resources to researchers on some sort of socialist (strict equality, disguised as "equity") basis, we are clearly not going to make the best use of our current stars.[13]

The second difficulty is how to get researchers and other knowledge workers aligned with corporate objectives while still giving them enough freedom to foster innovation. This involved finding the appropriate level of management for knowledge workers (Fig 11.4). At the right, we have the extreme of *micromanagement*, meaning continually looking over knowledge workers' shoulders; this will almost certainly suffocate their natural innovative impulses. At the left, we have the other extreme, *abdication*, the view that "these people are really smart so we should just let them do whatever they want." The chances of "whatever they want" being aligned with company strategies and goals are pretty thin.[14] The best management style—a happy medium—is *delegation*, where a large measure

---

[13] There are several other difficulties with the socialist approach. First: The stars, knowing their worth, will tend to leave for some venue that does not use a socialist approach. Second: There will no longer be any disincentive to prevent the untalented researchers—dare we call them "black holes" (for resources)?—from assailing the gates of the research department for employment, since they are assured that they will suffer no disadvantage with respect to their more able colleagues. Many of these assailants will be unsuccessful, but not all. Some of this success will be due to the always-impressive impulse in some managers to blur the distinction between *equity*—which is supposed to mean fairness (a very important but very subjective idea)—and *equality* (which is a simple mathematical idea, but quite irrelevant to the creation of high-performance teams). It does not take an experienced systems analyst to know that when the weakest people keep staying and the strongest people keep leaving (to be replaced by more weak people), the resulting team is not going to win the Super Bowl.

[14] The author once encountered a relatively small high-tech company in which the researchers were permitted to do pretty much whatever they pleased, while the purchase of pencils was carefully accounted for. This is a perfect example of over-managing physical assets while under-managing intellectual assets. (*Hint:* The pencils cost a dime; the researchers cost $100,000/year.)

of freedom is given within broad guidelines, subject to periodic reviews (at least once per quarter).

Figure 11.4: Find the Optimal Management Intensity.

| Abdication | Delegation | Micro-Management |

An old-fashioned military-like command-and-control management style does not work well with knowledge workers. They prefer to be *persuaded* and to have input to their work decisions and directions. This is, once again, where management by *leadership* comes in. To the extent possible, the manager should use *influence*, not *authority*.

According to Jack Welch, the legendary CEO of General Electric, another characteristic of great leaders is that they *face reality*. Indeed, they actively seek it. In Welch's words, "reality is rarely what it used to be, or what you thought it would be, or what you hoped it would be. Then, having found reality, devise a plan of action and act quickly." Welch also frequently asked himself five important questions about his own business leadership (see accompanying box).

Jack Welch's Five Questions
(for business leaders to ask themselves)

1. Do I come to work every morning filled with enthusiasm for the day ahead?
2. Do I reach my desk each day with an open mind, fully prepared to re-write my agenda, if necessary?
3. Do I energize my staff?
4. Do I have the drive, courage and conviction of a truly effective leader?
5. Do I execute well and deliver the results that I and my business strive for?

11.3 THE ENGINEER → MANAGEMENT TRANSITION: A CHECKLIST

The transition from engineering to management is full of challenge; the range of skills required to be successful in the management role is considerably larger than for engineering. In an engineering position, work is generally of a technical nature and requires some sense of engineering expertise, skill, or knowledge to obtain the desired result. Management, however, relies more heavily on advanced interpersonal skills to understand, manage and deal with various types of human behavior. These skills include, but are not limited to: **(a)** effective listening; **(b)** delivering performance feedback and coaching; **(c)** performance management; **(d)** dealing with difficult people; and **(e)** orchestrating team dynamics.

A Personal Experience

Neeraj Ghai reflects on his personal experience:

> During my time as manager of 21 staff, I found myself somewhat overwhelmed by the heavy reliance on my interpersonal skills and the problems and complications that arose when one of those skills didn't shine as it should have, if even for a second. When dealing with human interactions—and distinct from engineering work, which is characterized by technical certainties—I found myself confronted by hard-to-grasp complexities arising from the diversity of talents, needs, goals, styles and cultures of my staff members. All of these interactions, in addition to meeting my own work objectives and helping my team reach theirs, made management difficult.
>
> I learned my lesson early: Navigating the seas of managerial responsibility was far from easy, and required a heightened sense of human relations, including communicating, motivating and managing employees, meetings and time. I soon discovered that human relationships were the glue that held everything else together.

I quickly came to another realization as well: differential equations weren't going to help me. I should have taken one more Human Resources (HR) or Organizational Behavior (OB) course while in school.

The traditional activities in effective management—planning, budgeting, organizing, staffing, controlling and problem solving—are no longer sufficient as one attains the higher levels of management, especially with knowledge workers. More senior management has grown to include the distinct yet complimentary function of leadership, which hinges on one's ability to set directions, motivate, inspire and align people with the company's goals. Therefore, before one attempts the transition to management, it is important to reflect on the key leadership characteristics and on whether one actually possesses them. Indeed, this is the primary purpose of this chapter. Leadership and leadership skills continue to gain prominence as more and more organizations adopt the philosophy of perpetual organizational change.

The most important thing to remember is this: Managers must realize that their people come first; their own success is contingent on that of their people. Some of the key drivers of their success will be the extent to which they value, develop, motivate and celebrate members of their staff. To do this, managers and leaders must master the arts of communication, fine-tune their coaching skills, deliver constructive feedback and be comfortable managing people up or out.

The Six Key Characteristics Of Effective Leadership

We now examine (Fig 11.5) six key characteristics of effective leadership (the inner ring) and identify a set of leadership competencies for each characteristic (the outer ring). By examining these characteristics and associated competencies, readers will be able to identify both their current strengths and the areas they may have to develop further should they wish to transition into management.

Figure 11.5: Key Leadership Characteristics and Associated Competencies.

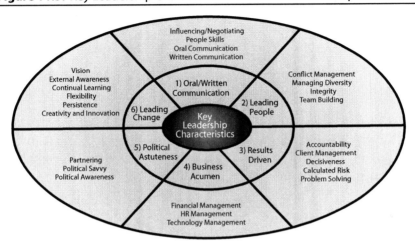

These characteristics and competencies are not simply binary traits that one either does or doesn't have—they are all well-known and well-defined skills that one can develop (although some are more difficult to learn than others). Effective leaders have learned, practiced and honed skills that may formerly have been shaky. We now examine these skills in more detail to permit readers to gain a better appreciation of what these skills entail and to assess their own developmental opportunities.

Leadership Characteristic 1: Verbal (Oral + Written) Communication
No skill is required more pervasively than that of communication. In order to be successful, business leaders[15] must possess advanced *oral* communication skills (meetings, conversations, presentations, telecons, press briefings, etc.) and *written* communication skills (memos, letters, reports, e-mail, etc.).

Effective oral communication is a two-way channel. It isn't just about talking; it also involves effective listening. Ask yourself:

---
[15] As should be obvious from the context, we are speaking here of business and other organizational leaders—not of athletic leaders or battlefield leaders, for example.

When someone is speaking to you, how often do you have a response ready to go before the other person finishes talking? (If so, that's a bad sign.) Are you even *listening* to them, beyond the point of response formulation?

Are you comfortable running meetings? Making presentations? Speaking in public? Do you have the skills and confidence to negotiate over the phone or face-to-face? Do you have the communication skills necessary to influence people that have no formal reporting relationship to you? Engineers pondering the management option should commit time to reflecting candidly on these and similar questions relating to the strength of their communication skills. Effective managers and leaders should be able to communicate smoothly with clients, with staff and with all levels of management. They must also listen actively, provide constructive feedback when appropriate (are you comfortable doing so?), and express complex ideas clearly in both written and oral format (do you enjoy all forms of public speaking, both formal and informal?).

A personal example from Neeraj Ghai:

> Several weeks ago, I had a conversation with one of the seniors at my company. He told me that I had an attitude problem, based on his having been made aware of a situation in which (he said) I was "belly-aching" about one of the many options from which I had to choose. The problem arose one day when I went to dinner with another person at the company and he took what I said the wrong way (and also chose to tell my boss). As I tried to defend myself to my boss, I was told that the reality or even my intentions didn't matter; only my audience's perception did. I had to admit that this was true.
>
> With that admission, I reflected on how important communication really is and on the care that should be devoted to every word one chooses, how one constructs verbal assertions, the visual images and body language that accompanies the words—and, to the extent possible, the perspective of the recipient. Each of these components together manufactures the total message. In this example, I hadn't dealt with all these aspects successfully and therefore I ended up being misunderstood.

At the management level, being misunderstood can be dangerous. Effective communication is critical to lubricating difficult situations, to playing the political game when necessary (see Chapter 6), to avoiding being misunderstood, to building relationships across the organization that facilitate the accomplishment of results, and to fostering an environment of open and clear communication. Well-chosen words and thoughtful responses can influence one's way to even further success. Asserting one's ideas with confidence and enthusiasm means that they will likely be understood and taken seriously.

It is difficult to over-state the importance of communication. If you lack well-polished communication skills or are not a "people person," a transition into management is probably not your best option at this time. For many, communication may be the most time-consuming skill set to develop, but it is also the most important.

Leadership Characteristic 2: Leading People
Leading people includes managing diversity, demonstrating integrity, building trust and forming a capable and high-performing team, while managing conflict situations as they arise. Let us now examine each of these in a little more detail.

Managing Diversity and Demonstrating Care. It all starts and ends with people. Leaders therefore need to demonstrate that they care about their people. They must empathize with their needs, their concerns and their goals. (A good indication of this personal characteristic is to ask staff members whether they believe that you show such empathy.)

In addition, given today's multicultural environment, especially in larger metropolitan cities, effective leadership involves an appreciation for diversity. This involves embracing the value of diversity in people, including their culture, gender, age and interests. Leaders must learn about other cultures and be able to motivate people from different backgrounds. They must recognize the value of diverse opinions and views while helping others to do the same.

In management, one must develop a sense of comfort in dealing with and overcoming the tensions and complexities that tend to accompany diversity—and yet make sound decisions. This means managing the environment to ensure that systems and policies are equitable, acting in a manner that is inclusive, respectful and non-judgmental, and dealing with ethical situations immediately and proactively. Leaders who demonstrate such understanding, care and ability will enjoy a greater commitment from the people they work with, leading to higher levels of success.

Demonstrating Integrity and Building Trust. As a result of the Enron, WorldCom and similar fiascos, ethical businesses are now placing greater emphasis on preaching, teaching and instilling integrity across all layers of management and staff.[16] At the management level, the demonstration of integrity has taken on new meaning and higher priority. Effective leaders must be able to demonstrate honest, ethical behavior in all their interactions. They must be aware of the policies and procedures, controls and compliance—and everything else in between.

Any individual thinking about transitioning to management should first feel comfortable with his or her ability to consistently treat others fairly and respectfully, to encourage others to do the same, and to confront actions or behaviors that are borderline unethical. Integrity requires the demonstration of consistency between values and behavior. Leaders, after all, must lead by example and live out the values of the organization.

Team Building. Leaders in all walks of life must be people oriented and possess the ability to treat other managers as peers, not competitors. They must unite their teams by utilizing resources well, including human resources. Managers must be able to work with their teams, to coach them and to manage their performance. It also helps if they have the gift for relieving tension in uncomfortable situations through the use of humor. This means provid-

---

[16]See the discussion of ethics in Chapter 13.

ing feedback and guidance as appropriate to help others in their development, identifying skill gaps and leveraging training and mentoring resources to close those gaps. The judicious shaping of roles and assignments can leverage the strengths and develop the capabilities of team members.

In fact, coaching may be one of the most underrated aspects of a career in management. Day in and day out, managers must give their employees the benefit of their experience and expertise and encourage them to share their experience and insights as well. Coaching improves and develops people (as opposed to punishing them). Leaders must work to create the conditions that enable subordinates to excel.

Hand in hand with effective communication is a leader's ability to motivate, inspire and influence. Effective leaders foster and instill a sense of energy, excitement, ownership and personal commitment in others. A leader who is in tune with the needs and interests of team members (and even those that have no direct reporting relationship) can engage and motivate others and inspire action by addressing these items.

Conflict Management. Effective leaders must guide their teams through situations of turmoil and confusion, requiring a sense of calmness and level-headedness. In any team environment, a leader must be savvy enough to prevent potential problem situations and to minimize negative impacts arising from various confrontations. Leaders must demonstrate comfort and confidence in their ability to deal with, manage and resolve conflicts and disagreements in a positive and constructive manner.

Leadership Characteristic 3: Being Results Driven (Plus)

In the past, an individual's performance rating was largely based on what they accomplished. Many organizations today are placing more emphasis on the "how" component of performance. An effective leader must demonstrate a drive for results that brings together elements of influence, initiative and decisiveness. Leaders must be able to obtain the results they desire through others by leveraging strong influence relationships, while still allowing them to preserve these relationships for future collaboration.

In many situations, a manager may not have the appropriate authority or expertise to leverage support and needs to be creative and resourceful. In addition, a drive for results is often contingent on one's ability to demonstrate a bias for action (taking the initiative). Effective leaders proactively determine what needs to be done, contrive a game plan for attacking the task—and, most importantly, execute it. In order to accomplish this, one must demonstrate decisiveness—which in turn relies on one's ability to separate necessary facts from unnecessary information, on one's rational judgment and on one's confidence to act on that judgment.

All in all, leaders should be able to maintain a positive, focused attitude in pursuing a goal, despite obstacles that may arise along the way. Their teams will perform at higher standard levels if they are able to foster a sense of urgency in others for achieving goals, to instill a strong sense of commitment and confidence in others, and proactively to break down barriers.

Effective leaders possess an inherent drive for results, which means that they hold themselves and others accountable for rules, responsibilities and performance. Managers are required to monitor, evaluate and measure the attainment of desired outcomes (on time and within budget). In addition, effective leaders are able to balance client relationships and readily adjust to shifting priorities and demands. At all times, leaders and managers must demonstrate good judgment by making thoughtful decisions, even in the face of ambiguity and imperfect information.

Achieving desired results will often involve taking deliberate and calculated risks that couple logic with instinct. Leaders must be comfortable with hazards and not be paralyzed by risk aversion (Chapter 7). In addition, managers should understand how their unit fits into the larger organizational picture; they will organize short-term tasks to complement and support long-term priorities, as set out by the strategic business plan.

Leadership Characteristic 4: Business Acumen

Effective leadership requires one to be able to demonstrate skill and comfort in the integration of financial management (funding, budgets, cost-benefit thinking, managing and monitoring expenditures, cost-effectiveness, etc.), human resource management (current and future staffing needs, and the selection, development and utilization of staff) and technology management (integration and understanding of technologies that enhance processes or minimize costs). Managers must see the bigger picture.

Leadership Characteristic 5: Political Astuteness

There is nothing as frustrating as a career that falls into the abyss because of a political comment made about you to someone of a higher rank. For example, someone could lose out on a promotion simply because of a comment made about him during a lunch between two people a year prior. To avoid falling victim to corporate politics (Chapter 6), one needs to be politically astute.

Effective leaders are able to diagram the (real) power structure of their organization and to identify those individuals in the organization that will support them and their ideas. They are able to build partnerships, effective alliances and a network of relationships that help them get things done. To be politically astute is to be aware of the political environment, to understand people, to understand their hot buttons, and to have an overall sense of the power dynamics of the business. Effective leaders listen carefully to the concerns of powerful people and departments and know where to turn for support and resources should they need them.

Leadership Characteristic 6: Leading Change

Leaders must know where they wish to lead. They must be guided by a vision. Leadership involves constantly articulating that vision to others so that they can see the path they must travel to achieve those desires. The effective leader knows whom to see when the group needs outside help; they have a strong external awareness.

Persistence and strategic thinking enable them to pursue—and achieve—desired objectives. Through patience, commitment and hard work, leaders are able to accomplish change and march closer to the shared vision for the betterment of the organization. In short, they can execute the strategic plan.

To lead change successfully also requires one to be *comfortable with ambiguity*. An effective leader must demonstrate flexibility when needed. In fact, he must strive in the face of ambiguity, see change as an opportunity, and encourage creativity and innovation from those around him. In the end, a leader must be able to translate creative ideas into business results. In order to work effectively in the face of ambiguity, shifting priorities and rapid change, a leader must successfully exhibit certain competency behaviors.

First, one must take personal responsibility for leading by example when supporting organizational change efforts. Second, one must help others adapt to change by communicating change efforts to them, framing them so as not to appear threatening, and helping those most affected by change to overcome inhibitions and fears that may adversely affect their performance. Third, it is important to deal constructively with mistakes and setbacks along the path of change in order to maintain a positive outlook in difficult situations.

Over the long haul, it is important to remember that organizations and teams must be agile and innovative. They must remain change-seeking and not get complacent or overconfident based on past success. A leader who continually looks for new ideas for positive change even while focusing on the implementation of current change programs will tend to be more successful in his career than one who does not.

In the face of such continuous change, it follows that leaders must exhibit the desire to *continuously learn*, while demonstrating flexibility and adaptability. Effective leaders adapt quickly to new job responsibilities and are good at juggling many different projects or tasks simultaneously, without losing sight of the staff whose individual contributions they manage.

11.4 A FINAL WORD...

Finally, this word of advice: If you are making the transition from engineering to management, you are in effect promising to invest in your ongoing personal development. This means that you are excited about seeking out and interacting with people who have strengths and knowledge that you do not possess. In order to further facilitate personal and career growth, you must seek out and actively participate in professional betterment courses. However, even this is not enough. You must then apply those lessons to your professional activity. All too often managers attend training courses because they feel they should, but walk away with nothing more than a few training manuals, a couple of models for more effective management and a notepad of scribbled information that will never again be looked at. All too often, what is learned is not practiced and what is not practiced is not improved.

As much as leaders need to put forth a conscious effort to improve their management capabilities, they also need to be realistic about what pathway they should take given the status of their current skill set. Roles need to fit people just as much as people need to fit roles. Clearly, before making important career decisions, one must understand these roles.

The leadership role heavily relies on one's ability to network, set the group agenda, think conceptually, analyze unstructured situations and interpret partial or convoluted information in pursuit of answers. This often leads to feelings of being out of control, or of being too constrained or lost in ambiguous definitions of success in a world of interdependency. This framework is virtually the opposite of the control, competence, narrow technical focus, independence and concrete definition of success that most engineers are familiar with. To transfer into a management or leadership role from an engineering role requires that one not only be comfortable with and prepared for these dramatic changes, but to view them as important opportunities.

CHAPTER 12

Entrepreneurship

A Business is Born

Most companies start their lives much as humans do—full of promise, much loved by their parents, but with a largely unpredictable future. Not all human babies thrive. Nor do all new businesses. As stands to reason, the degree to which newborns—human or organizational—are successful depends in large part on the skills of their parents. The purpose of this chapter is to talk about corporate infancy: how to decide whether to start a company; how to start one; how to nourish one; and ultimately how, with some notable exceptions, a parent must make the difficult withdrawal as these companies—through their various developmental phases—take their place in the adult corporate environment.

If all kinds of new companies are included, from well-conceived startups to little Mom-and-Pop businesses with little formal preparation, the chances of survival are not all that good, as shown in Figure 12.1. However, for those with more education and understanding of business, and with adequate advice and financial support, the chances are much better. It is also hard to overstate the importance of startups in our economy and culture. Virtually all the net new jobs come from startups—the biggest businesses having gotten, by definition, about as large as they are going to get. Statistics show that about half of all innovations (and 95% of *radical* innovations) come from new and smaller firms.

Figure 12.1: Survival Rate for Startups (all kinds).

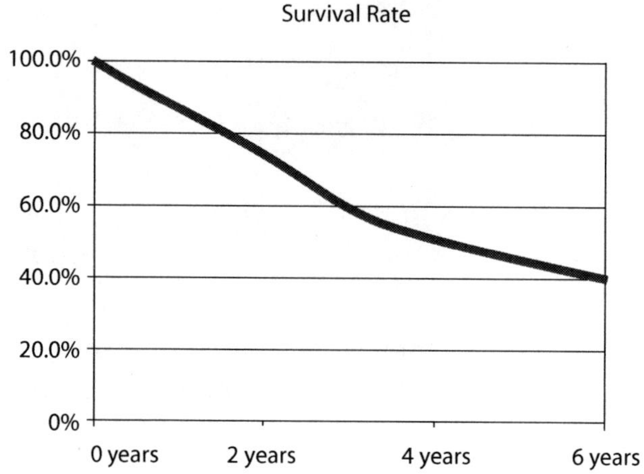

Startups Under Discussion

The newborn businesses given the spotlight in this chapter have two essential characteristics: **(a)** They are founded[1] by engineers, scientists or others with a similar background as practitioners of applied science (or by a larger group in which such individuals are prominently featured as key personnel), and **(b)** they are intended to provide the nucleus for a growing company.

With respect to **(a)**, no inference should be taken that other companies, where engineering or other science-based technology drivers are absent, are not important. (They are just not the subject in this book.) Further, with respect to **(b)**, companies that are not growth-oriented are almost always *lifestyle companies*, meaning that the founders (and principal owners) have decided that they will use

---

[1] This definition comprises many disciplines, including—in addition to engineers, technologists and physical scientists—medical scientists, bio-scientists, other life sciences and all their derivatives, pharmacy, dentistry and a plenitude of other examples.

a corporate vehicle as their personal source of employment. While this model often serves these owners reasonably well, the unfortunate other employees of these somewhat stunted businesses usually get squished when they venture into the business domain of larger carnivores.[2] In any case, if a business does not regard growth[3] as a good thing, it suffers from an illness analogous to some of the unfortunate genetic illnesses suffered by human babies. The owners don't want all that much, and don't get all that much, but the non-owner employees[4] get even less.

Connections With Earlier Chapters
There are many connections between the present discussion and the material of earlier chapters, among them these:

- Connections with Chapter 3 [Engineering School]: Students often come across exciting new technologies in engineering school, especially those they help to develop through their own studies. These can be commercialized.

- Connections with Chapter 4 [The Long Path]: For one familiar with the historical flow—from fundamentals like abstract maths, applied maths, science, and engineering maths to the ultimate betterment of the human condition through technology development and commercialization—startups can be especially satisfying. To one who is the intellectual driving force (founder) behind such a startup, and especially if the startup

---

[2] There is no reason for outsiders to invest in such a business, since its purpose is to benefit the owner-employees, not the owners generally.
[3] Growth is a form of change. Yet, as discussed in §7.2, "most humans are averse to change." This makes the *sincere* desire for continued growth, among founders of new companies, less common than may at first appear.
[4] A special warning should be given about one sort of small, lifestyle company: If one is considering working for a small company that is a family-owned company, this means that *nepotism*, not *meritocracy*, will almost certainly be company policy. Favoritism will be shown to relatives or close friends of those in power; they will be given jobs, inflated salaries, or other perquisites. For such companies all other strategic considerations are a smokescreen. Hail the nearest taxicab and find a real company to work for.

proves to be sustainable and successful, the founder can see this company as a legacy contribution to human betterment.

- Connections with Chapter 5 [Marketing Concepts]: There is always the danger that a technology-based company will fail to see the importance of a sophisticated relationship to a well-defined market. Indeed, Chapter 5 was intended as an antidote to such fatally narrow viewpoints.

- Connections with Chapter 6 [Office Politics]: Interpersonal friction should be less frequent in a very small company only just formed by a very few like-minded people. However, it may not take long . . .

- Connections with Chapter 7 [Risk Taking]: Some have claimed that entrepreneurs like to take risks, but we shall see below that this is not the case.

- Connections with Chapter 8 [Accounting]: The services of an accountant will be required to prepare annual financial statements that satisfy the tax department and the company's bank.

- Connections with Chapter 9 [Innovation]: Starting a new company is, by itself, an innovation. Still, one must have something further that is novel and attractive to offer the market if the company is to be successful.

- Connections with Chapter 10 [Intellectual Capital]: A startup has much less intellectual capital than do larger companies.[5] However, it frequently has more IC than any other type of capital! It is not unusual for a startup to have less than $10,000 in

---

[5] This discussion does not encompass new subsidiaries set up by existing companies to exploit a new technology, product or service. Such startups will be well-funded and will have many IC assets.

tangible assets. This explains, for example, why investors in new companies are very interested in patents and other intellectual property—the most tangible component of IC at this early stage. (See Fig 10.5.)

- Connections with Chapter 11 [Leadership]: Entrepreneurship is a form of leadership. Indeed, leadership is a necessary (and almost sufficient) condition to be successful as an entrepreneur.

Other, less obvious connections will also be made as our discussion of entrepreneurship proceeds.

Entrepreneurship vis-à-vis Innovation

There are many connections between entrepreneurship and innovation, so it is not surprising that they can often be reasonably confused. An author, on the other hand, needs to use a variety of words with precision to make plain his intended meaning.

It is artificial to think of a new business that doesn't have some strong element of innovation. As pointed out in the subsection of §9.1, titled "Innovation is Everyone's Business," we certainly do not imply here that innovation is always technical innovation; the innovation that prompts a new company to be formed may be of many types. (Nevertheless, it would not be surprising if most readers of the present book were most excited by a new company based on a technological innovation.) Thus, to a large extent, the word *entrepreneur* implies a strong context of *innovation* without the need for an explicit statement of this fact.

At the other end of the spectrum, very large companies can, without the stimulus of change, become stultified, heading for mummified. What they need is *innovation*—something sturdily new that will carry them forward. Such innovation within a LargeCo often requires personal characteristics from an individual—often called a *champion*—who is completely committed to the innovation and knows how to make it happen. Champions have much in common with entrepreneurs, sharing several vital common characteristics. Some have dubbed these champions *intrapreneurs*, a witty word that embodies a great truth.

Despite these interesting cross-meanings and insights, we shall restrain ourselves to using *entrepreneurship* for startups and very small companies, and to using *innovation* for larger companies.

12.1 SIZE DOES MATTER

We could begin the development here by examining either **(a)** what an *entrepreneur* is, or **(b)** what a *startup* is. In other words, we could initiate the discussion by focusing on either the creator or the created. We shall choose the latter but the picture will be incomplete until both aspects have been addressed (in §§12.1 & 12.2).

Companies Have Phases Of Growth

The analogy between the growth of companies and the growth of humans, while obviously less than perfect, is appealing because the latter is so well understood by all. For example, we have all heard many times that children are not just miniature adults.[6] It is equally true that small companies are not just miniature large companies. There are stages to company growth, just as there are to human growth, and an understanding of these stages is important.

Before commenting further on these company phases, however, perhaps a note of caution is in order regarding the precision that can be given to assertions on this subject. One speaks of the "hard" sciences, meaning subjects like physics and chemistry where universal laws have been discovered and expressed with mathematical precision. Then one speaks of the "soft" sciences, a domain where general laws and precise quantitative relationships are rare or absent and one must make do with second-best treatments, including empirical

---

[6]Modern neurology continues to make elegant contributions to our understanding of the growth of the human mind and the functions of the human brain, from the prenatal stage to old age.

rules, schools of thought and well-intended studies (often with conflicting results). It is (or should be) the intent of researchers in these softer sciences to harden them through the development of more unassailable truths (via the scientific method), thus leading to a deeper understanding and more confident predictions of future behavior.

Still, the soft sciences are almost brittle compared to the gelatinous material from which the proto-theory of company growth phases is erected. Many authors have described these supposed phases. Some refer to four principal phases; some to five, or six, or more. Companies are so variable in their characteristics and so complex in their details that it is little wonder that there is not a more substantive theory available. What really matters is to understand this: Companies that grow—the only kind of interest here—do not do so continuously, uniformly and homologously. They grow by fits and starts, sometimes retreating but more often advancing. The former problems of smallness wane in memory as the new problems associated with largeness loom in importance.

In fact, given the clear absence of any persuasive general theory of business, one might wonder how businesses navigate at all. The answer is simple enough: Corporations are not on an autopilot that has been designed based on some elegant system of algorithms, each based, in turn, on an elaborate theory. They are instead organically and finely controlled by the myriad decisions made every day by their employees. This is, indeed, the global effect of the intellectual capital spoken of in Chapter 10—and explains why that chapter was subtitled, Operating System for a Network of Brains.

Examples: Communications And Financial Resources

Shown in Fig 12.2 are two examples of potential company problems that differ markedly depending on company size. Everyone would recognize that effective communications and adequate financial resources are two important attributes of any company of any size. Yet the degree to which these issues are generally burdensome tends to be heavily dependent on company size.

Figure 12.2: Two Examples of Large vs. Small Company Problems.

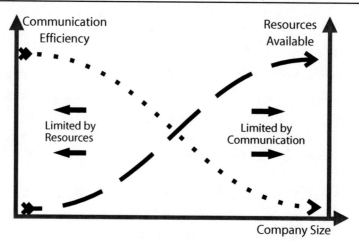

To take communications first, large companies have many impediments[7] to successful intra-company information. Certainly, technology has helped greatly with some dimensions of this predicament, but others remain stubbornly intractable. For small companies, in contrast, communication is relatively straightforward. With startups (as the limiting case), Jane can just shout across the office, "Dick, are those projections ready yet?"

At the other extreme, consider financial resources. Once we admit that no company of any size has or ever will have enough cash to do all the things that it wishes to do, and once we recognize that even large corporations can become financially stressed, it must still be accepted that, as a rule, SmallCos will have Balance Sheets that are weaker than those of LargeCos.

A Three-stage Growth Model

Let us begin with a relatively simple (yet revealing) model of company growth. In this model, there are three basic stages: Newborn,

---

[7]For example, large companies tend to have many buildings (and even many locations), making face-to-face dialog difficult; the 80% of non-verbal communication available when meeting in person becomes more rare and more expensive. A HugeCo also has more layers of management, each a filter on effective communication. HugeCos also have many more professional specialties, each a potential "silo" identified by a special lexicon and culture—and nourished by vocophilia.

Rapid Growth and Market Leader. In human terms, this corresponds roughly to babies, teenagers and adults.

Three of the key dimensions are shown in Fig 12.3, corresponding to the strategic, financial and operational challenges of each of the three company sizes. Most of the words in Fig 12.3 are self-explanatory, so there seems little need to exposit at length. Some of the more complex ideas have been discussed earlier (e.g., intellectual capital in Chapter 10) and others will be addressed later in this chapter (e.g., angel investing and venture capital).

Figure 12.3: A Three-Stage Company Growth Model.

While Fig 12.3 is better than a complete fog—or the unrealistic assumption that businesses of all sizes are basically alike—it is still quite truncated as a model for the concerns of growing businesses. Accordingly, we shall now briefly examine the progression of growth using a model that reveals more detail.

A Six-stage Growth Model

A business growth model that has finer granularity[8] than the three-stage model is a six-stage model developed by the Ontario govern-

---

[8] It can be assumed here that the challenges of company growth—whether depicted as three stages, or six, or some other number—are largely similar throughout the western world. Certainly there are, within various cultures and jurisdictions, important distinctions between markets and critical differences in legal details. However, at the general level being considered here, the CEOs of similarly-sized companies (from, say, Huntsville, Alabama, or Esbjerg, Denmark, or Coquitlam, British Columbia) will benefit from sharing their mutual experiences on many more stories of achievement and defeat than the CEOs of companies of widely different sizes, even though the latter are from the same geographical localities.

ment. In fact, the granularity is even finer than that, because these six stages cover[9] just Newborn to Rapid Growth, to use the terminology of Fig 12.3.

The salient characteristics of each of these six stages are (much abridged!) as follows:

Stage 1. One-Man Band. (The old idiom is inclusively corrected, of course, to one-person band.) In this the earliest stage, the founder is king, still owning about 100% of the shares and making about 100% of the decisions. The market offerings are very narrow, planning is rare and exports are the exception. The focus is on the product, not the people: there are no job descriptions, no employee training, and little management structure. The main challenge is access to capital.

Stage 2. Early Success. Another doubling (or more) of revenue has occurred. A significant market connection has been made and the fragrance of potential success has been detected. Line extensions and adjacent markets are developed, but this success makes the business more complicated, so at least one new manager is hired—an embryonic management structure! The founder must delegate[10] some responsibility. The founder also starts some intense self-directed study, with a strong functional emphasis on perceived weaknesses (finance, marketing, etc.). Still no long-term thinking, though, and the primary sources of advice tend to be customers and suppliers (in that order).

Stage 3. Focus on Basic Operations. Another doubling (or more) of revenue has occurred. The focus now is on providing and optimizing the operational infrastructure to support the successful products. Management structure is further formalized: a new layer is added, possibly through a key promotion, and authority and responsibility

---

[9]See D Rumball, "The Six Stages of Growth," Publication No. 3 in the *Leading Growth Firms Series*, published by the Ontario Ministry of Economic Development and Trade, 2001. This publication summarizes the results of an extensive symposium held among CEOs of Ontario's leading growth firms, defined as companies employing between 20 and 500 people with at least 50% revenue growth (year over year).

[10]Abdication and micromanagement are the two opposite (and extreme) dangers here; see Fig 11.4.

are further distributed. Management training is further emphasized (especially for the founder). More private investment is brought in, broadening the shareholder base. Still no long-term or strategic thinking, nor consideration of alternate strategies. Very customer-focused.

Stage 4. People Crunch. Another doubling (or more) of revenue has occurred. For the first time, the founder realizes that, to grow further, it's really about key people. While now attempting to master most or all of the functional issues of the business, the founder realizes that not all the internal managers can handle the demands of continued growth. Outside talent is seriously sought. Some large corporations are sometimes brought in as partners (possibly with outright sale of the company). Strategic thinking begins in earnest and products and markets are more finely tuned to the best growth prospects.

Stage 5. Professionalizing. Another doubling (or more) of revenue has occurred. More outside managers, who have performed well in much larger and more complex organizations, are brought in. Although concerns regarding how the larger company will be managed are thus being addressed, working with these independent professionals requires some adjustment on the part of the founder(s). The founder's personal ownership tends to further decrease in percentage terms, though it increases in share value.

Stage 6. Mature Corporation. Another doubling (or more) of revenue has occurred. To quote directly from the primary reference (Rumball, *op cit.*), "In Stage 6, leading growth firms become diversified corporations, led by strong management teams with great depth and the ability . . . to sustain long-term growth. They have multiple product lines and markets. Most such companies are exporters [to many countries]. Their CEOs are leaders rather than managers, presiding over a multi-layered management team with embedded processes of planning and decision-making."

What could be more exciting for a founder than to parent a wonderful corporation? Yet, there are many opportunities, in the above generic company life story, where the founder (or the small founding team) can go badly off the tracks.

Many Opportunities To Foul It Up

We shall discuss the entrepreneur more personally in the next section (§12.2), but it is already clear from a thoughtful perusal of the six-stage narrative above that there are a great many ways for a founder to go astray. Progress through the six stages may either be stultified (if the entrepreneur is slow to recognize the challenges of each new stage) or even aborted (if the entrepreneur fails completely to navigate one of the stages).

In fact, the exposition just given of the various successful changes and adjustments required at each stage of a growing, thriving business also contains, by implication, a recipe for a long list of ways the founder can go wrong. There are two very general things[11] a founding entrepreneur must understand in order to conduct the symphony of growth: **(a)** his or her own nature, and **(b)** the nature of the business. Plainly, it is the objective comparison of **(a)** with **(b)** that is required—another needed talent. One must have the attitude of objective self-examination. Such a comparison requires rationality and humility, the latter perhaps not always plentiful among the traits of a true entrepreneur.

Further, the six-stage entrepreneurial scenario above was only briefly sketched—and only generically, without the many more particular pitfalls that lurk in specific businesses. It is a rare entrepreneur who can successfully navigate the entire gamut of vulnerabilities, but the more one learns, both about the needs of the business and about one's own strengths and weaknesses, the more a fatal limitation will be kept from undermining the success of the business.

12.2 THE ESSENTIAL SPECIES: ENTREPRENEUR

Although *entrepreneur* is a word familiar to most, its meaning is sometimes misunderstood. Some guidance is available from the

---

[11] A little good luck doesn't hurt either!

fact that it is a French word, based in part on the preposition *entre* (meaning "between") and in part on the verb *prendre* (meaning "to take"). Thus, an entrepreneur is one who takes resources as needed from several places and combines them to form a new business. The metaphor of solving a jigsaw puzzle may be helpful, except that the entrepreneur frequently has to make the pieces as well.[12]

How Do Entrepreneurs Think?
Further light is shed by these less etymological but more practical comments:

- An entrepreneur is someone who conceives solutions to problems and then launches a company to implement that solution.

- Entrepreneurship is creating and building a company of value from practically nothing.

- To an entrepreneur, problems are opportunities—markets looking for a product or service.

- The entrepreneur may not be the inventor or the ultimate manager; he or she bridges the no-man's land in between.

- An entrepreneur does not do what everybody else does.

One of the great myths about entrepreneurs is that they like to take risks. Not at all; it is wanton gamblers who like to take risks and they are rarely successful. Instead, and with reference to the risk-

---

[12]Chapter 1 in G Kawasaki, *The Art of the Start* (Portfolio, 2004) discusses the reasons an entrepreneur may wish to create a new company. This is an excellent book and is "must reading" for any would-be entrepreneur. Other helpful books include: JA Timmons, *New Venture Creation* (Rev. Fourth Edition), Irwin, 1994; HH Stevenson et al., *New Business Ventures and the Entrepreneur*, Irwin 1994; WA Sahlman, HH Stevenson, *The Entrepreneurial Venture*, Harvard Business School Practice of Management Series, 1999.

reward relationship shown in Figs. 7.7–7.9, entrepreneurs accept carefully calculated risks, including large ones *if* the high reward (also calculated with care) justifies the high risk. In other words, they operate much higher up the risk-reward curve, but they are very much *on the curve* if they are to be successful.

Entrepreneurs ask themselves the following sorts of questions before they undertake a new venture:

- **Is my idea good enough?** They do as much market research as they can afford.

- **Do I have the management skills needed?** The six-stage scenario in the last section illustrates this principle. If key skills are missing, they team up with someone who provides them.

- **Do I have the personal qualities needed?** These include confidence, leadership, a desire to turn a profit, and gritty persistence.

- **Can I get the financial resources I need?** Good ideas can starve for want of cash. See §12.3.

Entrepreneurs who are engineers (or have some other high-tech background) must also ask themselves the following all-important question:

> **Paramount Question for Entrepreneur Wannabes Who Are Engineers**
>
> Am I ***market*** driven or am I ***technology*** driven?

Unless one is market driven, one should not attempt to be an entrepreneur. It is not enough just to talk the talk; one must also walk the walk. It is fine (it may even be wonderful) to be technology *based*, i.e., to be a high-technology company; but one must be *driven* by an interest in customers and one must be hyper-responsive to the market.

There are several ways of determining an honest answer to the above paramount question for any engineer. Here are two:

Way 1. Call a meeting with some of your engineering pals and say that the purpose of the meeting is to discuss the possible market for [some technical idea or product]. Then sit back and notice how the participants behave.[13] Some will stay on topic, chatting about various market aspects; others will lapse into long-winded technical exchanges about the magnetic this or the software that, considerations that are related to the stated purpose of the meeting only tenuously, if at all. When the market is being seriously discussed in detail, some will be highly focused and genuinely interested, while others will start to glaze over, looking wanly at the table and waiting until this "marketing stuff" stops.

Way 2. Conduct a strategy review in your organization. Key marketing and technical people should be present. Everyone knows that the company has two important technologies (one developed internally and the other licensed in), which enable a product to be manufactured and sold to a market, as shown:

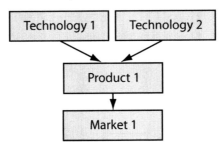

The question is: How do we grow the company (in addition to the obvious effort to grow the existing market)? The engineers suggest the strategy shown below, pointing out that, by developing a third technology, they can, through combination with Technology 2, create a second product that can also be sold to their existing

---

[13] As the reader may well have guessed, the author has been involved, inadvertently, in many such experiments.

customers. "See," they say, "We're market driven. We don't just want to develop new technologies; we also want to address the market possibilities."

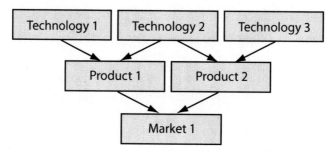

"I'm afraid I see this as being technology driven," says the director of sales and marketing. "Given a choice between developing another technology and developing another market, you will almost always choose to develop a new technology. I think there are several potential markets we could develop with our existing two technologies."

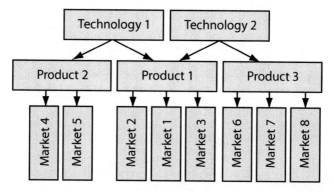

He then sketches the diagram above and uses it to provide a detailed roadmap of seven additional markets that could be tackled, complete with a new product offering.

One of the engineers muses to himself, as he gazes bleakly at the table, "This market stuff never seems to go away."

There are times, of course, when new technologies are needed, but the above parable does make a point.

Chapter 12: Entrepreneurship

Can Engineering Students Be Entrepreneurs?

With rare exceptions, students or the just-graduated are too inexperienced to start a company all by themselves.[14] A student is not an entrepreneur just because he doesn't want to work for anybody, or because he can't find a job, or because he wants to make a lot of money—fast! Having a small committed *team* and a strong business idea, however, raises the odds that a startup will be successful.

The urge to start a new company often comes during one's graduate research program, but it occasionally happens as early as the senior undergraduate years. All such students should realize that, if they have participated in the development of a commercializable entity, they have a clear intellectual property (IP) position that must be recognized. This means that, even though most professors are scrupulously ethical (the author is unaware of a single instance where a professor has stolen intellectual property from a student), students should leap to the nearest outside advisor if a research supervisor suggests any connection between giving up IP rights and the award of a university degree.

At the same time, a graduate student who helps to develop something that can be commercialized must recognize that, since the supervisor (professor) suggested the research idea, raised the money to fund it, made laboratory or computer assets available, continually advised the student during the research, gave courses that developed the student's understanding, helped to write the research reports, suggested that there might well be a commercial possibility, made key contacts inside or outside the university with respect to the commercial options, used his or her professional reputation to lend credibility to the solidity of the research results at all stages of the process, the student's entitlement to the total award is considerably less than he may initially think. It's not about time

---

[14]For most of his career, the author would have set the probability of success of a student startup as dauntingly low, but the students he encountered while Director of the Jeffrey Skoll BASc/MBA Program have demonstrated that there is a non-negligible set of students whose gifts encompass both engineering and business, with distinctively high quality in both. For these, entrepreneurship at a young age is often possible.

spent; otherwise the custodian and the bus-driver would be in for a big chunk as well.

Three Common Entrepreneurial Weaknesses

There is no shortage of pitfalls and pratfalls on the way to building a strong company, but three are so common and so important that they deserve special mention.

The first is the *refusal to delegate*. This takes us back to Fig 11.4. Most entrepreneurs have a specialty of their own and have built the first beginnings of their business on that expertise. As the company grows, it is difficult to entrust most of that functional area to others—to avoid, in short, the temptation to micro-manage. Some would-be business starters try to do too much themselves, thus paralyzing the firm and stunting its growth. Perversely, they often take pride in the 60 hours/week they work, somehow not realizing, as a real entrepreneur would, that under their incapable leadership the company will never double in size because that would require them to work 120 hours/week (which of course they cannot). The business quickly asymptotes to the micro-size that is consistent with the boss doing all sorts of things himself that should really be delegated to someone else, thus freeing up time for the really critical areas.

The second common entrepreneurial error is the *refusal to outsource*. Many excuses are given. Here are two: **(1)** "It costs too much to pay someone outside to do this," and **(2)** "Why let someone else make a profit on this? We'll do it in-house and keep the profit ourselves." Both these excuses would be anathema to a real entrepreneur who would realize, instinctively, with respect to Excuse 1, that it will cost even more to do it internally,[15] and, with respect to Excuse 2, that there are such things in business as core competencies so that, unless the subject activity falls within the *core competencies*

---

[15]Especially if the person who does this activity internally is the owner, who should be looking to offload relative trivia to focus on strategic issues and whose relatively high compensation guarantees that it costs considerably more to do this work in-house.

of the firm, paying a fair margin to an outside organization will produce a result of higher quality and lower cost than the do-it-all-yourself crowd can bring themselves to realize.[16]

Still, even with an intelligent approach to outsourcing, the employees of small firms will find themselves doing a wider range of activities than at a large company; this is another fundamental difference between LargeCos and SmallCos. In fact, this broader territory of activities for each employee is why some people prefer working for smaller companies.

The third common error made by many self-labeled entrepreneurs is the *refusal to exchange some control for cash* to fund growth. There will be a fuller discussion of venture capital and related topics in §12.3, but for now it will suffice to point out that when new investors put V cash into a company that was worth V immediately before the investment, the new (book) value immediately after the investment is $V + V$. This means that, while the ownership of the founders is reduced in percentage terms—a process called *dilution*—the number and value of their shares remain unchanged. Thus, owners cannot argue that their personal holdings are reduced in value by outside investment. What has changed, however, is the *degree of control*. It is the classical question: Would you rather be a relatively large fish in a small pond, or a relatively small fish in a large pond? With the synergies and faster growth that investment makes possible, entrepreneurs realize that, with an increased capital base, they can become a large fish in a large pond.

Can The Entrepreneur Scale?

One often hears about businesses *scaling* (and we shall get to that subject presently), but here we refer to the ability of the *entrepreneur* to scale. We know that the intent is to create a rapidly growing com-

---

[16]There are also some areas where outsourcing should be the *only* option considered. For example, attempting to get a corporation started without the benefit of competent legal counsel is the company equivalent of having a baby in the back seat of a taxicab, and, if forced to choose, the author's money would be on the cab driver to produce a generally better result.

pany—otherwise, this isn't entrepreneurship as meant here—yet we must recognize a very important fact: Unlike a large company, where the top people have risen through their merits, in the entrepreneurial company, the founder or founders started at the top and the business has grown around them. This raises the key question: Can these founders scale as the company grows? If not, the growth will grind to a halt, the best people will leave, and deterioration will set in.

The three common entrepreneurial errors described in the previous subsection provide three examples of such impediments to growth. A similarly motivated assessment has been given by Hamm,[17] who mentions four qualities that, paradoxically, are handy for the founder when the company is created, but become unwelcome constraints to growth if the founder cannot scale with (*grow with*) the company:

> **Impediment 1. Loyalty to Comrades.** Loyalty to individuals present at the founding is a positive personal trait on the part of the founder. Still, loyalty to the company (meaning to all its stakeholders) is clearly a higher priority for the company's CEO. As companies grow from inception to adulthood, there will always be those who cannot navigate from their success in the tiny environment of a startup to similar (scaled!) accomplishments in the vaster environment of a much larger company.[18] (In fact, this present discussion asks whether the startup CEO is a case in point!)
>
> **Impediment 2. Task Orientation.** This affliction is the reverse of attention deficit disorder. Everyone knows that focus is generally a good

---

[17] J Hamm, "Why Entrepreneurs Don't Scale," *Harvard Business Review*, Dec 2002, pp 2–7. (The detailed discussion here of these four constraints is the author's and should not be blamed on Hamm.)
[18] There can be a reverse phenomenon, of course, in which someone who is with a LargeCo decides to take on a more senior position in a SmallCo, and finds they cannot adapt. But often they can do wonders for the smaller company based on their broader experience—and here lies a great irony: Suppose a person with much experience and success in a LargeCo is being considered for a senior position in a recent startup, with the aim of bringing it to a new and higher level. One of the SmallCo founders objects to this hire on the grounds that "This person is used to a large company, so they may not understand a smaller company." A reasonable question, but if a founder objects on these grounds every time someone from a LargeCo is being considered, and on these grounds alone, it seems that this founder has, perhaps unintentionally, revealed that he or she is really a control freak—and has no intention of the SmallCo ever being anything but small!

idea, but this CEO takes it too far, concentrating on one challenge to the virtual exclusion of all others. This weakness is often closely related to the refusal to delegate mentioned above in the subsection on Three Common Entrepreneurial Weaknesses. Although this can happen with respect to any discipline of fundamental importance to the company—finance, technology, marketing, etc.—in this book we are particularly interested in founders who are engineers (or who have a similar background in science or high-tech) and who start a company based on a spinoff from that technical work. If this engineer thinks for one moment that he can simply find a commercializable item and then form a company with himself installed as CEO, all the while continuing to spend long hours in the company lab doing ever more engineering (or even some much lesser activity, such as bookkeeping), he is sadly mistaken.[19] Either he, or his newly created company, will fail.

Impediment 3. Single-Mindedness. This flaw is similar to the foregoing. Attention to one problem (whether it is in one's home field or not) or complete reliance on one's own views, while ignoring the opinions and advice of others who also have the best interests of the company in mind, is a recipe for failure.

Impediment 4. Working in Isolation. Many technical people are introverts. CEOs cannot be introverts. They cannot hide in their offices; they must interact with, and learn from, all the company stakeholders, including shareholders, investors, employees, suppliers and—most important—customers. In short, they must be leaders. They can do this only if they are "out there," mixing (including socially) and developing mature points of view.

So the question is: Can you grow with your company? Many of the necessary skills have been sketched in earlier chapters. Yes, some talents seem almost intrinsically personal endowments. One marvels and appreciates these gifts. But much can, and must, also be learned. Indeed, if one were asked to cite an excellent example

---

[19]There is nothing wrong with being committed both to one's personal contributions to engineering and to one's new company. The error is to pretend to be a company officer while at the same time behaving in a completely incompetent manner (as a company officer). Any founder-engineer who will not or can not act as a senior manager of the new company (much less the CEO) should, as a first priority, hire someone who can. It's that simple.

of the need for life-long learning, founder-CEOs would be strong candidates. If you become aware of a relevant talent, don't do the childish thing (resent it or ignore it), do the mature thing (try to acquire it). A founder-CEO should, at the end of every year, relish her personal growth achievements during the preceding year, attained through hard work, acute observation, and an iron will to improve her abilities. She may be assured that, by so doing, she is doing her best to foster the growth of her company as well.

Can The Business Scale?

Having considered the question of whether the entrepreneur can scale, we now ask this: Can the *business* scale? (Or, more accurately, can it *be scaled*?) As with many questions in microeconomics,[20] the explanation for present readers can be reduced from many pages and many diagrams to a few sentences, relying on the mathematical tools and concepts possessed by engineers. The really key idea here is *linearity*, referring to a systems approach to business modeling, wherein there are defined *inputs* and defined *outputs*. (Some earlier figures, such as Figs. 3.12, 10.8 and 10.9, exhibit this kind of system thinking for businesses.) The system can be said to be linear if each output is a linear combination (superposition) of the inputs.

For example, an investor's input *to* the business is his capital and his output *from* the business is his annual return. Or, as another important example, an input of interest to the CEO could be the annual revenue and the output of interest to the CEO could be the net earnings.

With the inputs and outputs thus identified, we can ask: Is the business linear? In the investor example, if the business is linear, twice the (absolute) investment will produce twice the (absolute) return. This is a reasonable assumption unless the investor is a very

---

[20]The reason for using the somewhat odd word *scale* (rather than, say, *grow*) is the implied reference to the principle in microeconomics known as the *economies of scale*; see, for example, RG Lipsey, *Economics*, HarperCollins, 2003. Another principle that seems to pop up with unwelcome frequency—but that pushes in the opposite direction—is the *law of diminishing returns*.

Chapter 12: Entrepreneurship

large one, meaning a pension fund or some other kind of relatively major investor, whose investments can actually affect the stock value and whose actions at an annual shareholders meeting can affect both stock value and company strategy.

For the CEO example (of greater interest to us here), does twice the revenue produce, *ceteris paribus*, twice the earnings? If so, the business is linear in that respect. Consider, however, the very simple business model discussed briefly in §8.7, wherein the profit $\pi(n) = (p - c_v)n - c_f$, where p, n, c_f and c_v are respectively the price at which each unit is sold, the number of units sold, the total fixed costs, and the variable cost (cost per unit). Evidently,

$$\frac{\partial \left(\frac{\pi}{n}\right)}{\partial n} = \frac{c_f}{n^2} > 0$$

which shows that the profit per unit grows with the number of units sold. Thus, in the sense used here, the business model is somewhat nonlinear; for example, a 10% increase in the number of units sold leads to more than a 10% increase in profit.[21]

We thus witness the celebrated economies of scale! This is a welcome nonlinearity for startups because it promises that the faster they grow, the faster they will grow in the future. Unfortunately, this result from *micro*economics, a consequence of fixed costs being smeared over more and more units as the business grows, and which predicts that the business will grow exponentially year over year, is clouded by **(a)** the fact that most fixed costs are not really all that fixed with respect to long times and large growth, and **(b)** further considerations from *macro*economics, which provides the usual realistic constraints that always restrict supposedly exponential growth.

---

[21] Some readers may feel more comfortable with an alternative, calculus-free derivation, in which the profit when the unit volume sold is increased by k% is compared with a k% increase in profit. The fact that $\pi[(1 + k)n] - (1 + k)\pi(n) = kc_f > 0$ again shows that the relative profit increase is larger than the relative sales volume increase.

To glance first at the maths, suppose that, instead of the simple cost model

$$c(n) = c_f + c_v n$$

we have, instead,

$$c(n) = c_f + c_v n + \varepsilon n^2 \quad (\varepsilon > 0)$$

indicating that there may be some cost components that increase more rapidly than would expected based only on sales volume. (We use the symbol ε to drive home the principle that this term may be relatively small at first, but that it will manifest itself and dominate the model as the business becomes ever larger.) The results just found above are now modified thus:

$$\frac{\partial\left(\frac{\pi}{n}\right)}{\partial n} = \frac{c_f}{n^2} - \varepsilon \quad \text{or} \quad \pi[(1+k)n] - (1+k)\pi(n) = kc_f - \varepsilon(1+k+k^2)n^2$$

depending on whether one is comfortable with calculus or prefers the calculus-free version. In either case, it is plain that at some point (i.e., at some business size, as represented by n), and even if ε is quite small, the ε term will push back on the economies of scale and will tend to limit the size of the business.[22]

Is this a real effect? Or is this just some abstract mathematical hypothesis? Very much the former. Figure 12.2 has already indicated the costly problems of communications for the company as it grows to more floors, in more buildings, at more sites, in more countries. There are many other saturation pressures as well. Larger companies, which usually become public, are subject to extensive and costly financial reporting requirements.[23] Furthermore, as com-

---

[22]That is, for n sufficiently large, the right-hand sides of these equations become *negative*, indicating shrinkage, not growth.
[23]In the U.S. these are mandated by the SEC—the Securities and Exchange Commission.

panies grow larger, they become subject to an enormous burden of duties, responsibilities, rules, laws and costs that are mandated by our modern legislative bodies. Although small companies are generally spared from most of these myriad costly items, they become burdensome for mature corporations. Many other examples could be cited.

Still, this panorama of escalating costs for ever-larger businesses misses an even more fundamental set of limitations that have nothing to do with legislation or social policy. Any trained engineer with mathematical sensibilities will already have spotted this problem several paragraphs ago. The simple fact is that unlimited growth is always an unrealistic figment of theories that work for variables within some perimeter of applicability but that ignore the larger scale factors that lurk on the boundaries of applicability of such theories.[24] Neither General Motors, nor IBM, nor General Electric, nor Walmart, nor Microsoft, nor Intel has ever beaten the limits to growth—although they have done better at it than most—and so long as they operate in a market economy, they never will.

First of all, there is a finite market. No matter how much a given company dominates its market, and regardless of how successful that company is at expanding that market, the market nevertheless remains finite. Then there is the little matter of competition: thus far, we have taken the microeconomics view that we examine the best moves for one company within a static environment. But the

---

[24]Every day, engineers face similar limitations to the theories they must use. As discussed in Chapter 4, engineers do not have the luxury that abstract mathematicians do—of idealizing their artificial world of consideration, in bizarre and drastic ways, so that their results can be claimed to be exact. When, for example, a young engineer attends her first class in the first week of first year and is told that she should think about a mechanical spring that will, in response to any force F, extend a distance x calculable from $x = F/k$, where k is some constant characteristic of the spring, she should say something like this to herself: "Hello? If this spring is pulled sufficiently far, will it not break? So much for k! Or, if the spring is compressed more and more, won't things get messy quite rapidly? It would take a thousand k's to represent the chaos of that crush. I'm willing to use this simple linear spring model for small deflections, but for large displacements? Forget it!" Engineers get used to being skeptical about the limits of simple models that seem to work within some bounded scope; they would therefore not likely be surprised that the so-called economies of scale or any other similar exponential growth tendency does not explain all the horizons on the economic landscape. Engineers—especially the applied science variety—have an uncanny ability to detect when some allegedly quantitative approach will work and when it won't.

environment is anything but "static." There are, in all the really interesting markets, several other players, and their strategies always include killing your company as collateral damage. Without stout offense and defense from your company, this all leaves the pretty idea of economies of scale as just a few equations on a piece of paper.

Markets as a whole cannot expand faster than the rate of growth of the economy and the latter is carefully shepherded by a central bank in most if not all of the virtually free-market economies. This constrains the average growth possible for any economic sector and for business generally. With globalization, the basic principles do not change all that much, but the level of vulnerability to competition of any rapidly growing business must now be calculated, not just within some local jurisdiction, but worldwide. Moving to a world stage with one's business—which one should plan eventually to do[25] with one's startup—means that the opportunities are greater, but so is the competition.

12.3 ACQUIRING MUCH-NEEDED CASH

The execution of virtually every good idea for a small company (or for any company) involves the expenditure of cash, yet cash is in very short supply for startups. Cash is the oxygen that fuels company growth, but it will not be available unless the new company can convince those who possess such cash that the return will likely be sufficient to justify their risk—and unless, in addition, the leaders of the startup have the charisma to convince potential investors that theirs is a better investment than the large clutter of other startup alternatives.

Stages Of Investment

A typical investment story will now be painted, starting with a primitive startup and ending with going public (the Initial Public

---

[25]Recall Fig 12.3 and Stage 6 (§12.1) in the "six-stage model of growth to maturity."

Offering, or IPO). The same caveats apply here as in §12.1, when stages of growth were considered—the primary admonition being that the stories of no two businesses are exactly the same. Unlike, say, butterflies, which begin life as an egg, then hatch into a caterpillar (or *larva*), then form a chrysalis (or *pupa*), and finally mature into a beautiful butterfly, no such predictive exactitude is available for companies.

Figure 12.4: Possible Stages of Future Investment for a Startup.

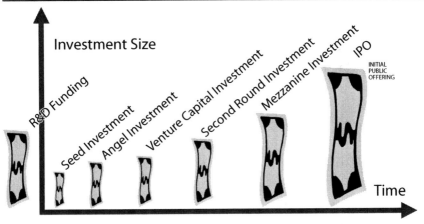

Nevertheless, the lay of the land is shown in Fig 12.4. It is unlikely that each and every stage will happen for any particular company, but all these stages for raising capital are potentially available. Shown at the outset in Fig 12.4—for $t < 0$, before the formation of the startup—is an R&D component of capitalization. Although not usually of interest to business theorists, all the pre-startup R&D expenditures (§9.3) should really be given some due credit if we are speaking primarily of high-tech startups, whose competitive edge is some form of technical innovation. This is not just a matter of giving credit where credit is due: if company management does not realize that similar major expenditures will also be needed in future for new product development (and perhaps also for some more basic research) it will be operating on a myth, not reality. The company will be a one-trick pony and will not be able to realize its full promise.

Seed Investment

As the name suggests, this is the earliest stage of capitalization for a startup. Notionally, it typically lies in the range $5K–$50K. If the entrepreneur has some spare cash of his own, this may be thrown in; the stories are legion of successful entrepreneurs who increased their house mortgages[26] eventually to become multimillionaires. Additional capital, in small but helpful amounts, may come[27] from family or friends.[28]

The important point is that this stage of external investment is enabled by one's *personal relationships*: the investors share either confidence (through personal knowledge) in the entrepreneur's ability to succeed or wish to support a respected friend or loved one in a worthy endeavor. It may be noted, however, that despite the relatively small size of investment, these seed investments[29] can lead to nontrivial ownership if the startup grows significantly.

Angel Investment

Although the seed investing just discussed is sometimes called angel investing for obvious reasons, we save this appellation for a somewhat different (and slightly later) class of investor. Angels do not initially have any connection with the entrepreneur; instead, they are wealthy individuals who have decided that there is money to be made or fun to be had (note the uniqueness of the latter moti-

---

[26]Entrepreneurs mortgaging their homes to start up their businesses may have led to the false myth that entrepreneurs like to take risks. However, the truth is that entrepreneurs will take risks, even large ones, if they rationally believe that the business rewards are so large that the risk is justified.

[27]Some have, only semi-jokingly, referred to such seed investment as 3F investment, referring to family, friends and fools.

[28]The art of finding money in strange places and avoiding unneeded expenditures for startups has been dubbed *bootstrap finance,* based on the idiom "pulling yourself up by your own bootstraps." See, for example, Bhide A, "Bootstrap Finance: The Art of Start-ups," *Harvard Business Review,* Nov–Dec 1992.

[29]Situations where someone just *gives* money to the entrepreneur, no strings attached, do happen, but are beyond the scope of the discussion here.

vation) through investing in very small firms. Angels rarely allocate a substantial fraction of their (high) net worth to angel activities, but it adds spice to their overall portfolio, and to their lives.

The risk at this stage is somewhat less (though still quite high) and the amount invested is typically in the range $50K–$500K. In the U.S., which has the most developed system in the world for capitalizing emerging businesses, angel investment is the single most important source of capital. There are between 100,000 and 500,000 angels currently active[30] in the U.S. Angels usually want revenue of $3M–$30M in five years, and annual growth of 15%–25% is attractive, although this would be too low for venture capitalists (VCs), as discussed next.

Another attractive characteristic of many angels is that they themselves are seasoned businesspeople, perhaps even successful entrepreneurs, who have made their money and are now interested in helping other startups. They are often willing to spend substantial lumps of their own time in rainmaking, utilizing their vast experience and exploiting their contacts to help[31] the fledgling company. Often, angels are willing to work in groups (formal or informal syndication), thus sharing the risk and multiplying the investment.

Venture Capital Investment

The next stage in capitalizing company growth is through *venture capital* organizations.[32] Note that these are organizations, not individuals, who are willing to invest, say, $0.5M–$5M in a still somewhat risky venture. The risk has gone down and the investment size has gone up, as shown in Fig 12.5. They demand that there

---

[30] U.S. angels invest $5B–$10B annually in 20,000–30,000 new U.S. companies.
[31] This provides an opportunity to observe one sure sign that a self-labeled entrepreneur is not really an entrepreneur at all, just a business loner: The advice and insights of a seasoned angel are regarded as interference. This startup egg may eventually become a larva, but it will never be a butterfly.
[32] Chapter 7 in Kawasaki's *Art of the Start (op. cit.)* is very helpful here. So, too, are A Rock, "Strategy vs. Tactics from a Venture Capitalist," *Harvard Business Review*, Nov–Dec 1987; R Quindlen, *Confessions of a Venture Capitalist*, Warner Books, 2000; and JW Bartlett, *Fundamentals of Venture Capital*, Madison Books, 1999.

be something genuinely innovative about the business proposition (new technology; or a new product, process or service; or a new marketing concept).

Figure 12.5: The Evolution of Risk Reduction and Investment Increase.

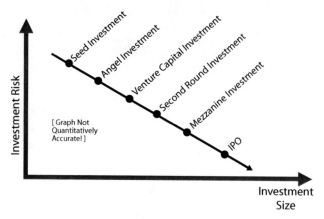

Some venture capitalists (who have a strong reason to be accurate and objective) have stated that they tend to place entrepreneurs into one of three categories:

Category 1. The Lifestyle Entrepreneur—You enjoy your own Co. Working for yourself is paramount. Continuing your pleasant existence is much more important to you than creating the next $1B Co.

Category 2. The Empire Builder—You are happy with the Co's growth rate and you love being the ruler. You wouldn't sell the Co if your life depended on it.

Category 3. The Serial Entrepreneur—You'll expand the Co to the best of your ability, sell it or go public, then start the cycle all over again.

These VCs know in less than five minutes into which of these three categories you fall and invest only if you are in Category 3.

VCs demand a significant (not necessarily controlling) participation in governance (a seat on the Board of Directors); all the components of a strong management team must be in place; and the new product or process must have passed (at least) the early prototype stage (with IP protected). Having some actual sales is often a clincher for many business sectors. Assuming technology readiness and market readiness, the biggest need for funds is working capital to penetrate—or perhaps even create—the market. VCs often talk about *exit*[33] *strategies*, meaning how to get their money out in (say) five years.

VCs typically want 35% or more annual return on their money. They tend to impose onerous restrictions and requirements on the founders (in addition to the obvious ownership dilution) but these must be traded off against the other significant benefits that VCs can bring, in terms of working capital, sage advice, catalyzing contacts, etc. In any case, by this stage a simple love affair between the founder and his technology (the neat technology trap) is far from sufficient for success; indeed, it is not even necessary.

Further Stages Of Investment

Only a few comments will be made here on the further possible stages of investment, the details having become too case-specific and detailed for the purposes of this book. VC investment may be in more than one round, each subsequent round visiting further dilution on the existing shareholders, but being based on a much-enhanced valuation of the business.

Intermediaries can sometimes help with the process of hooking up with VCs (or angels). They typically charge 2%–5% of the deal, which is a major expenditure, and some agents with rolodexes of angels and VCs also want some stock as part of their compensation, which can be an even larger reward if the company is really going somewhere. Note also that these agents are different in important ways from, say, real estate agents. The latter offer three things: **(a)** access to the market, **(b)** presentational skills, and **(c)** negotiation. Angel and VC interme-

---

[33]Exit strategies should also be on the mind of the founding entrepreneur(s), although this is often absent for amateur entrepreneurs (literally, entrepreneurs for the love of it). Such strategies include **(a)** being acquired by a larger company, **(b)** second round or mezzanine financing, or **(c)** an eventual IPO. One cynic has waggishly remarked that share certificates in a small, private, illiquid company are useful only for papering over the cracks in the cottage outhouse.

diaries offer only the first of these—a promise to provide contact with angels and VCs. The presentation must be made by the entrepreneur (what investor wants some canned presentation from a third party in place of a convincing pitch from the key people themselves?), and the negotiation is also done primarily by the entrepreneur (it is assumed that, if she can't sell the company, she can't sell, period.) Shaky revenue projections usually lead to zero investment.

Often,[34] *going public* (technically known as the Initial Public Offering, or IPO for short) is viewed as a mature stage in the growth process. It is attractive to the founders because it enables them finally to liquidate their previously illiquid shareholdings. A large fraction of the populations of western democracies can participate in the economic activity of their countries, either directly by owning stock in companies they choose or indirectly through mutual funds, retirement funds, etc. The public stock exchanges make an enormous contribution to the economic component of democracy.

However, an IPO is a very large step for a corporation to take, and will have many implications (financial, reporting, and otherwise) every year thereafter. The best advice should be sought on how (or whether) to go ahead. The large brokerages, banks, legal firms and accounting firms have all the information needed.[35]

General Stance On Outside Investment

It has been assumed in this section that outside investment is always good. The implication has been that if one can get it, one should take it—and the more the better. Let's pause a moment and examine this assumption. One of the three entrepreneurial weaknesses described in §12.2 was the unwillingness to exchange some control for cash. Evidently, it is implied that one should never refuse to consider outside investment. Most entrepreneurs do spend

---

[34]However, some large, important companies do remain private, and, at the other extreme, even quite small companies can go public on smaller, more risky, "over the counter" stock exchanges.
[35]For example, the accounting firm KPMG puts out a helpful 100-page booklet called, simply, *Going Public*.

a major effort courting outside investment. Is this realistic? Upon what principle(s) is this appetite based?

We can start with accounting. It has already been remarked in §12.2 that, from a Balance Sheet perspective, when new investors put ΔV of cash into a company that was worth V immediately before the investment, the new value immediately after the investment is $V + \Delta V$. However, this is the increase in the *book* value, and if this were the only consequence of the investment, and the new cash were left as cash, there would be no advantage to the whole exercise. In fact, the reluctant entrepreneurs would be correct: they would lose some voting control (due to a dilution in their share percentage) with no other present or future change in the value in their shares traceable solely to the investment.

However, given the underlying assumptions of the accounting process, valuation of a startup by looking solely at book value is peculiar and inappropriate. Accounting is by nature rearward-viewing, which makes it singularly inadequate to valuate emerging companies whose primary source of value lies in future operations and growth. According to strict accounting concepts, new investors add nothing for pre-investment stockholders at the instant of investment. They contribute an additional x% to the value of the assets (in cash), and as a consequence, they now own x% of the company. How does this benefit the pre-investment shareholders? The accounting answer seems plain: It doesn't! Everything increases by x%, but precisely x% of that increased benefit accrues to the new investors, so where is the upside for old shareholders?

If the startup is one that just lumbers along; without innovation; with no vision other than the employment of its founder and his friends; with no (real) strategy other than muddling along in a lethargy of changelessness; then there is scant basis for accepting (much less seeking) external[36] investment. For young corporations,

---

[36] Nobody will realize this more quickly than these self-same outside investors. The problem is self-resolving because they will not invest.

in contrast, that are committed to rapid growth and that know how to achieve their strategic goals, outside investment makes sense.

To unravel this conundrum, we must return to one of the fundamental laws of microeconomics, the economies of scale. As discussed in §12.2, a company that is $x\%$ larger can do more than $x\%$ more and will increase earnings by more than $x\%$. The book value is almost irrelevant. A significant increase in capitalization permits the hiring of a key CEO; or enables the attraction of the marketing VP who will make all the difference; or will show the bank that this company has a rosy future, meriting a larger credit line or a more attractive interest rate; or will demonstrate to key customers that this company will be here a decade from now; or will prompt less stressful credit terms from strategic suppliers. The list is endless.

Thus, even though the arithmetic of simple financial statements is incapable of predicting its truth, the empirical fact is that any rapidly growing company will benefit from—and should seek—external investment, and this fact is ignored at an entrepreneur's peril.[37] In fact, one of the criticisms of the custodians of highly successful rapidly growing companies is that they sometimes leave too much new investment money on the table.

---

[37]Paradoxically, the only time this argument may break down occurs when an emerging company is so extraordinarily profitable that all its truly attractive business plans can be funded from earnings.

CHAPTER 13

Governance

Overseeing Management

One the most fundamental purposes of this book—indeed, the primary reason engineers may wish to buy and read it—is to promote the viewpoint that engineering, as a profession, makes an exceptionally fine foundation for management.

By management, we do not mean work responsibilities that, after paying one's dues for a few years, are best described as supervising some technicians. We also do not mean scenarios where, after a few more years of experience, the job description refers obliquely to administering a team that involves a few engineers but where the administration leaves little room for intelligence, judgment, or innovation. Supervising simply means that there is someone even newer and younger than the supervisor and that they require some attention; and administering usually implies that there is a thick policy manual that should be applied to the cases at hand as they come along. Not much scope for innovation here; not much opportunity for leadership either. In short, these are not the sorts of management of interest to us.

On the contrary, this book suggests, with neither exaggeration nor apology, that engineers, properly selected and groomed, are capable of the highest levels of management—*fully-fledged management, at **all** levels*.

This includes the most senior management in the company—the officers of the corporation and the members of the Board of Directors. This book therefore concludes with a chapter on these leadership roles. Among the many spires of human excellence at this level there are a few individuals whose first degree was in engineering, but the author has no doubt but that there could, and should, be a great many more. Practitioners in law, finance, marketing and accounting—and sundry other home disciplines—are commonplace in the boardroom. This is exactly as it should be: Each of these professional specialties has a critical role to play in the guidance of a major corporation. Is there room for a wider representation of disciplines?

With all the emphasis nowadays on technology—permeating most modern product offerings and lubricating the high-productivity internal processes in modern business—engineers who have the interest, training, energy and (most especially) the talent for top management positions should be sought out early and groomed to be corporate leaders.

Some whose formal training was originally in engineering leave their home organization, perhaps in frustration, to found their own enterprises (Chapter 12). Some of these are successful, some not—but one must wonder how many of these stories feature someone who would have welcomed more responsibility, authority and accountability in the organization he or she left.

Note carefully that any argument made here for having more engineers as CEOs or for more engineers in the boardroom is most emphatically *not an entitlement argument*. It is *not* claimed that there is some higher authority that would prescribe a more visible representation from the engineering professions. It is *not* alleged that engineering as a profession has some right to any preconceived allocation of spaces at various high levels of organizational power. Indeed, the entire political language of privilege from class, or, in the present case, privilege from profession, is anathema to the stance being advocated.

Chapter 13: Governance

13.1 ENGINEERS IN SENIOR MANAGEMENT

It is very important to maintain a sense of balance here, and to focus precisely on what is, and what is not, being argued. Should engineers have some kind of management rights by virtue of their being trained as, or later having experience as, engineers, and should they be given some sort of quota preference over other professions with respect to leadership positions? Of course not. At the other extreme, engineers should not suffer some intrinsic impediment to ascending the management ladder, nor should there be an invisible "nerd ceiling" above which organizational culture[1] prevents them from being entrusted with the leadership of important human organizational groups of whatever kind.

The profession of engineering should not be an inherent obstacle to the highest rungs on the leadership ladder; it should be a natural springboard. This basic cultural truth is not booted about; indeed it is obscured by contrary effusions of all kinds, in many arenas, and from many professions. While it is not surprising to observe that architects, say, do not expend a great deal of energy elucidating the many attractive management qualities[2] of engineers—to do otherwise would violate the basic principle of vocophilia—what is truly surprising (at least to one observer, the author) is that engineers do not do so either!

Senior Rungs On The Management Ladder

Figure 13.1 shows the top few levels of management within a corporation. It should be born in mind that organizational charts come in

---

[1] The author, through many conferences, short courses, professional societies, government associations and similar inter-professional gabfests, has noted that engineers—and, even more emphatically, their professional first cousins, applied scientists—are, as a matter of received doctrine, often assumed to be intrinsically disqualified for business leadership and senior management based solely on their primary profession. This applies whether the context refers to achieving a more elevated position in their current organization or to a dominant role in the entrepreneurial company they themselves have founded. These opinions are sometimes expressed by people who, by objective standards of quality, would find themselves quickly unemployed in any fair talent competition with the engineers they disparage.

[2] The current setup is that architects hire engineers, as necessary, not the other way around. This sets up architects as the managers, and engineers as the managed. Congratulations to the architects. It could just as easily have been the other way around.

an almost infinite variety of diagrams, and hence this diagram seeks to identify the generic levels without giving any special job titles or reporting arrangements—other that those that are always present.

Figure 13.1: Generic Levels of Senior Corporate Management.

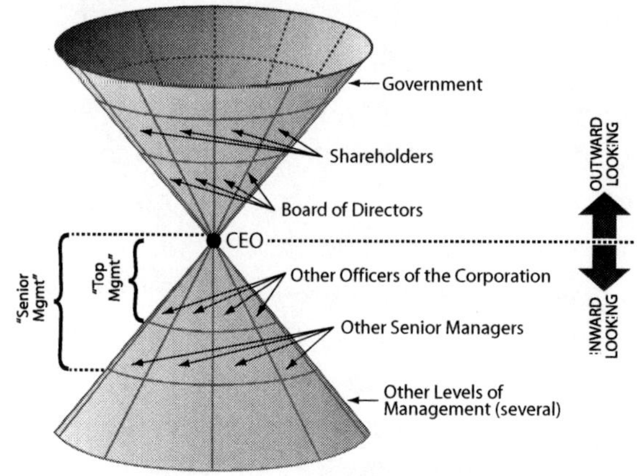

[Figure 13.1: A double-cone (hourglass) diagram showing, from top to bottom: Government, Shareholders, Board of Directors, CEO, Other Officers of the Corporation, Other Senior Managers, Other Levels of Management (several). Brackets on the left indicate "Senior Mgmt" and "Top Mgmt". Arrows on the right indicate OUTWARD LOOKING (above CEO line) and INWARD LOOKING (below).]

The lowest level shown—the highest level of Other Levels of Management, is still quite senior, especially in a large company. These would include positions like Chief Engineer, Technical Director, Director of Research and Development, etc., which would be natural career targets for engineers—but also a wide range of other positions.

Just above these positions are those who are just one level below the *Officer* level (defined in more detail below). These are the Other Senior Managers. There are fewer of these important positions than at the (many) other levels of management, but more than at the Officer level, labeled Other Officers of the Corporation in Fig 13.1. One Officer, the Chief Executive Officer (CEO), is so important that he or she gets a level of his or her own. The CEO and the other Officers will be discussed in more detail below.

By the time one attains the lofty position of CEO, one no longer has to report to a *single* person—indeed, there is no single person left to report to. The CEO does report to the Board of Directors (shortened, because of its frequent use in this chapter, to the acronym BoD). The BoD will also be discussed shortly, but one basic fact should be

Chapter 13: Governance

mentioned now: In addition to its other duties and activities, the BoD has primary hire-and-fire power over the CEO.

The Directors, in turn, are appointed by the shareholders, the owners of the company. Speaking, as we are here, about companies that are at least large and possibly very large, we can assume that there are at least thousands (perhaps many tens of thousands) of shareholders, while a BoD of more than 15 persons is unusual. (Banks seem to pad their boards with large numbers of Directors, and 20 Directors on a bank board would not make the Guinness Book of records.)

Finally, in Fig 13.1, we recognize that the most basic rules for how a corporation should behave are specified by the laws of the civil jurisdiction under which it was incorporated. In Canada, a corporation can be formed either under the laws of the national (federal) government or one of the ten provinces. In the USA, a corporation can be formed under the laws of one of the 50 states.

If the country involved is a democracy, so that the government can be thrown out or re-elected by the voters, this voting citizenry can be said to be exercising the ultimate power, at a level even higher than the Government shown in Fig 13.1. For example, if corporations are perceived by a majority of the voters as doing too much harm to the environment, this can become an election issue and a newly elected administration can enact newer, more environmentally friendly laws, which all corporations within that jurisdiction must observe and obey.

Officers Of The Corporation

To avoid repetition and to provide focus, it will be assumed here that the earlier chapters of this book have done justice to the Other Senior Managers level, and to the several lower Levels of Management depicted in Fig 13.1. More specifically, Chapter 11 on Leadership would pertain to these levels.[3] As a general example, using Fig

---

[3] Much of Chapter 11, particularly the material on leadership, also applies, of course, to the higher levels of management in Fig 13.1.

11.1 as a blueprint for how a typical professional engineering association views the careers of engineers, we might loosely associate Level E with Other Levels of Management and Level F with Other Senior Managers.

Before addressing the Officers directly, we digress for a moment to deal with a semantic issue that, left unaddressed, will undoubtedly lead to confusion. The term CEO is quite modern; the much older English word for the organizational leader is President (from the French, *président*, one who presides). Nowadays, the President of a company may not actually be the CEO; for example, he or she may be the Chief Operating Officer (COO), who[4] reports to the CEO. If, as is usually the case, the President is also the CEO, no doubt is left by the job title President and CEO. It is hoped that this redundancy will soon disappear from the corporate lexicon, along with its cause, the title President.

Figure 13.2: Old-Style Terminology for the "Direct Reports" to the Top Dog.

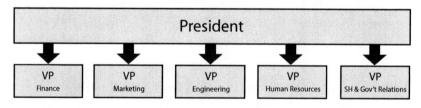

Associated with the title President are a list of Vice Presidents, all of whom report to the President. Five such persons are listed in Fig 13.2, but the number of Vice Presidents has no legal limit; some companies have organizational charts in which vice presidential positions proliferate to the point where one gains the impression that there is a good deal of title inflation going on. In many of the larger organizations, many sub-species of vice president have been created, including the following:

---

[4]More about COOs below.

- Senior Vice President
- Executive Vice President
- Senior Executive Vice President
- Group Vice President
- Divisional Vice President
- Regional Vice President

This staid old terminology is rapidly being replaced (at least in North America) by the Officer approach. This may be partly to suppress title inflation; partly because of the love of acronyms; or just for brevity. In any case, it does clearly specify who the Officers of the corporation are.

There are, again, many types and labels of Officers, depending on the nature and taste of the company. All companies, of course, have a CEO, whether that term is used or not; there has to be someone[5] at whose desk the buck finally stops. The CEO's management team comprises a group of Officers, each responsible for a major sector of the company's activities. Some possibilities include:

- Chief Financial Officer (CFO)
- Chief Marketing Officer (CMO) [e.g., Radware, Inc.]
- Chief Information Officer (CIO) [e.g., FedEx Corp.]
- Chief Research Officer (CRO) [e.g., AMR Research]
- Chief Technology Officer (CTO) [e.g., Nortel Networks Corp.]
- Chief Merchandising Officer (CMO) [e.g., Sears Holdings Corp.]
- Chief Investment Officer (CIO) [e.g., New Brunswick Investment Mgmt Corp.]
- Chief Administrative Officer (CAO) [e.g., New Boston Fund, Inc.]
- Chief Equity Officer (CEO) [e.g., New Boston Fund, Inc.]

The list is limited only by one's imagination. Note that the special needs of a particular business may well be represented in its choice of

---

[5]Occasionally one sees two Co-CEOs. This relationship rarely lasts very long. It is either false labeling or fundamentally unstable.

CXOs. For example, it is understandable that FedEx has a Chief Information Officer when the whole company runs on the data associated with the packages entrusted to it. Note also that the acronym versions can lead to a degree of uncertainty, there being, for example, two types of CIO and two types of CMO in the above list. The CTO position seems like a natural for engineers, but several others are also within reach.

People who are company Officers are expected to understand and speak for their companies overall, not just in their own area. This is the first level for which the claim to be domiciled in an organizational silo is completely unacceptable. Officers are truly *executives* because, as a team, they *execute* the company strategy and the directives of the BoD.

Officers are also empowered to sign on behalf of the company, a responsibility with serious legal implications. For those who are contemplating becoming an Officer of a corporation, it is a matter of the highest professional priority that they seek competent legal advice on their risks and responsibilities.

Chief Operating Officer (COO)

There is one rather special type of Officer (other than the CEO) who deserves separate mention—the Chief Operating Officer (COO). In this scheme, the COO is responsible primarily for operations within the company, so that the CEO can spend more time looking outward. See Fig 13.1 again; the top cone is unchanged although not explicitly shown in Fig 13.3 to save space. Sometimes the CEO and COO are nicknamed Mr Outside and Mr Inside, respectively.

Figure 13.3: The Role of COO in Senior Corporate Management.

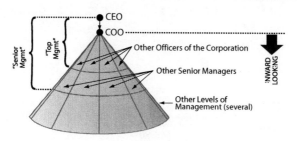

Chapter 13: Governance

In order for this to work effectively, the CEO-COO pair must work together without friction. It is important to clarify who is responsible on a day-to-day basis for what. (Ultimately, the CEO is responsible for everything.) Endless navel contemplation and sorting out of claims and counterclaims to power or responsibility wastes valuable time and will result in a change of assignments.

13.2 SERVING AS A DIRECTOR IN GOVERNANCE

Every corporation is required, by the legislation that gives it life and legitimacy, to have a body of individuals that *oversees* the management of the company. The concept of having a body of (largely part-time) individuals who oversee how well a particular organization is being managed is hardly unique to corporations or to corporate law. Universities and major teaching hospitals, for example, each have their Board of Governors or Board of Regents, as do (legitimate) charities and religious organizations. In each case, there is a Key Member of the Board who is responsible for both the day-to-day administration and the longer term strategic success of the institution in question. The Key Person (whether called a CEO, or a President, or a Principal, or other similar[6] name) must **(a)** be chosen, through some process, and **(b)** have his or her performance monitored.

The situation for corporations is as shown pictorially in Fig 13.4. The CEO reports to the BoD, which, in turn, represents the shareholders (owners). This process of managing at the strategic level without getting into the day-to-day minutiae of company transactions[7] is called *governance*, and is the primary subject of this chapter. Because large amounts of money are involved, providing opportunities both to do great good and great harm, a very

---

[6] In the U.K. and several other European countries, the term Managing Director is used for the CEO. This person is the one who sits on the BoD and it also the top manager in the company.

[7] In the words of WA Dimma (*Excellence in the Boardroom: Best Practices in Corporate Directorship*, Wiley, 2002), "a board should not, must not, indeed cannot manage. When it tries to do so, as it does from time to time, it is no longer a board. It has become a management."

sophisticated set of rules and procedures has been set up to accomplish this task.

Figure 13.4: The Corporate Governance Structure.

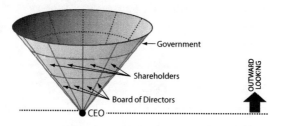

Moreover, these rules and laws are still evolving every year, quite rapidly, to accommodate technical advances and other innovations—and to keep one step ahead of the very small minority of senior managers who turn out to have the morals of bank robbers, though their weapons are not 357 magnums but phantom revenue entries, off-balance-sheet financial entities, stock-pumping schemes, miscellaneous money launderings and all manner of short cuts and end runs around ethical management practice. The details of these evolving rules also vary from jurisdiction to jurisdiction, so that only attorneys who specialize in securities law have any hope of keeping track of the latest wrinkles, loopholes and plugs.

Engineers On The BoD

Still, our purpose here is not to offer legal advice—far from it!—but broadly to describe corporate governance in a language that engineers can understand, so that they can decide whether they wish to set their sights on such a career role at some future[8] time. Considering the relevance of technology to the contemporary corporation,

---

[8]Most BoD members are not youngsters, and for very good reason (presently to become more apparent as their role is described). Engineers wishing to participate in BoD functions should think of this goal as an aspiration for the latter part of their career—unless either **(a)** they are on the BoD because they are a CEO (or other top manager) in the company, or **(b)** they are a founder or plurality shareholder of a startup company. In the vernacular, they must be recognized as either a Senior Statesmen or a Big Fish (relative to their pond).

and the outstanding foundation provided by a (strong, balanced) engineering education to almost every leadership role in the modern world,[9] it is perhaps something of a mystery why more engineers do not sit down at BoD meetings.

The author—as should be obvious by now—is of the opinion that the mystery, if indeed there is any mystery at all, is entirely cultural. Young engineers are taught in school to be great applied scientists (or perhaps to be great engineers), not to operate at even higher levels of responsibility. Engineers who work in organizations—including businesses—tend[10] to be encouraged to become even better engineers, not to provide competition to the managers who are giving this encouragement. These urgings to become even more proficient at one's home profession are entirely natural, and provide needed motivation and encouragement for the majority of engineers. Unfortunately, for an important minority—those whose destiny could be organizationally even higher—these limiting influences should be seen as cultural norms to be overcome. Once so seen, these seeming barriers can be surmounted without enormous difficulty—but only if they are recognized, internalized and assailed.

The Concept Of Governance

Let us spend a moment examining that strange but vital distinction between "managing" something, and "overseeing the management" of something. The latter is the essence of what is meant by *governance*. In Fig 11.4, the idea was broached that managers had a spectrum of management intensities at their disposal, and that a happy medium was needed between the extremes of *micromanagement* (managing every little thing, to the point where the person being managed gets the feeling that they are mere robots being con-

---

[9] For those engineers willing to recognize this fact, be inspired by it, build on it, work diligently toward it, and who have the natural ability to perform at these elevated levels.
[10] Engineers should surely prefer to work for companies that seek talent *everywhere* in their personnel—who encourage engineers (and all others) to create value in every mode that their interests, aptitudes, energy and preparation permits, whether in engineering in the classical professional sense, or as corporate leaders.

trolled by their supervisor) and *abdication* (managing so little as to be not managing at all, to the point where the person being managed gets the feeling that they are some kind of free spirit, not really accountable to anyone—and that, in particular, corporate goals are just obscure ideas that may relate to someone else, somewhere else, but certainly not to them). Finding the right point on a similar governance intensity scale is also important, as indicated in Fig 13.5.

Figure 13.5: The Fiduciary Role of Directors on Behalf of Shareholders.

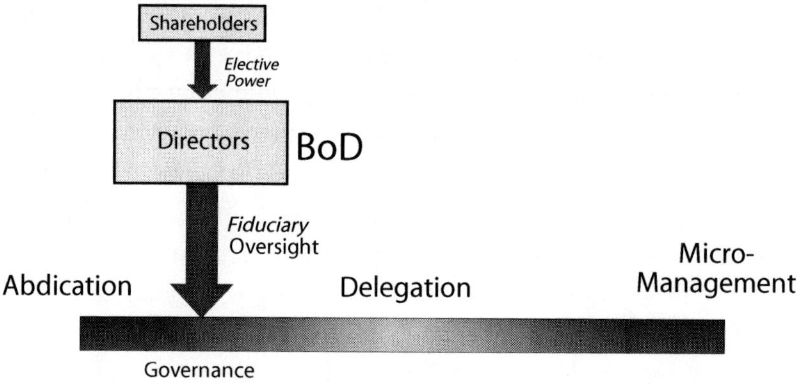

For Boards of Directors, there are two primary issues relating to governance style, as required by law:

1. The BoD cannot manage every little situation, yet must oversee the management of the corporation. Obviously, the BoD must concentrate its attention on only the most crucial issues, and the ones that are of long-term importance.

2. The BoD must manage the corporation in the best interests of the corporation. But what does this mean, exactly? One cleanly logical interpretation is that the BoD must manage in the best interests of the *owners* of the corporations—i.e., the shareholders. But there are more nuanced interpretations, some of which we shall discuss briefly near the end of §13.3.

This particular type of management—governance—is somewhat to the left of delegation in the management intensity chart shown in Fig 13.5, but is not abdication, especially in the modern era of heightened BoD responsibility.[11]

Fiduciary Behavior

Also shown in Fig 13.5 is the term *fiduciary oversight*, meaning oversight on behalf of the shareholders. This term is laden with a serious burden of legal implications, and it is impossible to discuss real-world Directors (and Boards of Directors) without reference to *legal liability*.

The concept of *fiduciary* is captured in the accompanying box. In legal terms, it is somewhat comparable to the concept of a *trustee*. However, a trustee has duties and obligations that are much more precise, whereas the fiduciary responsibilities of (and situations faced by) a Director are more varicolored and nuanced.

Fiduciary (Definition)

Adjective: *Of, or related to, a holding in trust for another.*

Noun: *One who stands in a special relation of trust, confidence, or responsibility in certain obligations to another.*

This makes the position of Director more interesting—and potentially more dangerous in terms of liability. Whereas a trustee has a very specific mandate—managing the assets of a wealthy widow may be the classic example—there is no end to the list of situations, lengthening almost daily, with respect to which a modern BoD may be asked to make fiduciary decisions. A fiduciary must act for the benefit of (specified) others, never for the benefit of himself. A Director must never be in a conflict (actual or potential) between his own interests and the interests of those he represents.

---

[11] For further comments on the distinction between management and governance, consult J Pound, "The Promise of the Governed Corporation," *Harvard Business Review*, Mar–Apr 1995.

Legal Liability Of Individual Directors

This is an area where both statute law and case law are changing[12] rapidly. We have thus far referred primarily to the BoD as a whole, and to Directors as members of the BoD. However, here there is a legal paradox: while Directors can act only[13] as a BoD, they are *individually* liable legally for certain BoD actions. This may dissuade some engineers from seeking such positions later on in their career, but in fact there are tools for handling this liability. The primary tool is to always be active—one might today say *pro*-active—in carrying out one's Directorial duties, and to do so conscientiously, personally, impartially and with complete integrity.[14]

A second tool for dealing with this liability issue is D&O Insurance, meaning indemnification, by the company, of its Directors and Officers against legal action by others. D&O insurance (a phenomenon that is only about three decades old) itself has to follow legal developments on D&O liability. In fact, there are even legal requirements regarding the obligation of the Co to indemnify D&O's! D&O insurance can be mentioned in the company bylaws, but it should be noted that such insurance is more difficult to obtain for a single Director.[15] The best overall insurance is that a Director always act in good faith and in the best interests of the corporation.

Normally, the most attractive Directors are individuals who have gained vast business experience and who have become financially content from their success. One might then ask: Why should they risk[16] this hard-won and well-deserved status to

---

[12]*Statute law* means law enacted by a legislative body; *case law* (or civil law, or common law) relies on legal precedents (earlier judicial decisions).

[13]That is, they cannot act individually on the corporation's behalf (except to the extent that they are also Officers of the corporation). As an important example, and unlike shareholders, Directors cannot vote by proxy.

[14]Translation: Racketeers, scam artists and those with feeble moral fiber or a dubious appreciation of ethics should not enter here.

[15]As observed by one legal writer [paraphrased]: The balance between Director action and Director indemnification, and between lawsuit and being found liable, is a (double) balance that does not and should not encourage (or even permit) behavior by Directors that is arguably and materially at odds with their fiduciary duties.

[16]Damages in many jurisdictions are based on restitution, not compensation. For example, if you sit on the BoD of a friend's company for $1 a year, and you are found liable (whether through intention or inadvertence) for $1M in damages, the fact that you were paid so nominally will not reduce your liability.

help with another company? (By believing a CFO's solemn assertions?) Obviously, these Directors are motivated by a desire to stay in the saddle, to spend their time in meaningful activities that they enjoy, and, quite probably and more altruistically, to give back to the business community some benefit from their own experience.

High-risk Behavior For Directors
It may be that some successful engineers, reading the above paragraphs, may think, "This sounds too risky for me. Who wants to place his head in the lion's mouth like that? I don't really even understand what the bad behavior areas are, so it will be harder to avoid them. I think it's more comfortable to stick to my slide rule, table of integrals and engineering handbooks and thus avoid this high-stakes Director Lottery."

This would be a shame. Engineers—only those, of course, with the right aptitudes, training and skills—are much needed in the boardroom, not because of some innate superiority, but because they are now sorely under-represented, considering the natural affinity between engineering and complex decision-making, and in view of the ever increasing correlation between strategic technology utilization and business success.

Thus, to move from obscure legalistic generalities to specific bad deeds, it can be said that one should stay away from the following actions:[17]

- Crimes (violations of criminal law).
- Misrepresentations (including material omissions) in a prospectus used to offer securities—for example, during an Initial Public Offering (IPO).

---

[17]Neither this list, nor anything else said in this book, should be construed by the reader as legal advice, which can only be obtained from a qualified lawyer or attorney. The idea is to provide a glimpse of the territory for career planning purposes, not to perform a microscopic examination.

- Mishandling of outside offers[18] to buy, or amalgamate with, the corporation.
- Breach of contractual obligations.
- Fraud.
- Negligence.
- Breach of confidence (using confidential information of a joint-venturer to make a profit).
- Insider trading.
- Diversion of a business opportunity.
- Competing against the Corporation.

A Unanimous Shareholder Agreement (USA) can connect many more dots between the Shareholders (as a group) and the Directors (as a group)—the ultimate embodiment of this pattern being that the Shareholders decide that they *themselves* wish to be the BoD, which does wonders for the fiduciary aspect of the arrangement, but may sometimes suffer a great deal from a lack of BoD quality.

Due Diligence

The legal phrase *due diligence* is often used in business and it certainly applies to Directors. It has a meaning completely consistent with the dictionary meaning of these two words, although there is an enormous overlay of legal interpretation in specific circumstances, and never more so than in the performance of one's BoD responsibilities. Diligence, as most people realize, means being earnest, persistent, steady, attentive and heedful. The descriptor "due" simply means "as appropriate in the circumstances."

The *duty of due diligence*[19] is a crucial component in a legal defense; e.g., if the Director can prove that he made such independent (!) investigations as enabled him to establish, on a balance of probabilities, that

---

[18] Here, if Directors are too malleable, they can be accused of selling too low or of leaving cash on the table. Yet, if they demand the highest theoretical price with great vigor, they can be cited by disappointed shareholders for rebuffing a beneficial sale.

[19] Shakespeare first used the phrase "due diligence" in *Pericles, Prince of Tyre*, in Act III (written in 1607-8).

these investigations were sufficient to give him reasonable grounds for believing that there were no misrepresentations, this can be used as a defense. In addition to the concept of the *diligent person*, the law also often refers to the *prudent person*, or the *reasonable person*. The quantity and quality of diligence, prudence or reasonableness (one less prone to -ness nouns might just say rationality) that is "due" will depend on the particular situation, and will ultimately be decided by a judge (if it comes to that). It will also depend on the natural expertise of the individual (e.g., accountant, engineer, etc.), what board committees he is on, etc. We might say, perhaps, that a *fiduciary* duty compels *active* steps on behalf of the shareholders, and that the *duty of care*, prevents *passive* (lazy) oversights, whether intentional or otherwise.

Directors with an engineering background might note that, just as Directors on the audit committee (for example) who are professional accountants will be held to a higher standard on financial matters than the average Director, matters involving engineering or technology should attract special attention from Directors with an engineering[20] background.

This is true a *fortiori* if the engineering specialty of the engineer-Director is pointblank in the area involved (e.g., an environmental engineer with respect to a violation of environment law, or a civil engineer with respect to liability for a collapsing bridge). More generally, what is prudent, diligent or reasonable is somewhat case-dependent (as is itself only prudent, diligent and reasonable).

Business Judgment Decisions

Generally speaking, courts are loath to second-guess actual business decisions of a BoD (or the BoD votes of a particular Director).

---

[20]In Canada (and perhaps elsewhere) the reliance on engineering extends further—to other Directors. Thus, any (non-engineer) Director is not liable for breach of fiduciary duty, the standard of care, or for many other sources of specific liability, if that Director has acted in good faith in reliance upon a report from an engineer, where the decision is highly reliant on the kind of appreciation for the subject matter that only a professional engineer could be expected to possess. (Not more so, however, than reliance upon financial statements which an Officer or the Auditor of the corporation represents to reflect fairly the financial condition of the corporation, or upon a report of a lawyer, accountant, appraiser or other person whose profession lends credibility to a statement made by such a person.)

Courts naturally focus more on improper procedure (conflicts of interest, breaches of fiduciary duty, lack of due diligence, etc.) in arriving at those decisions. There are at least four reasons for this reluctance:

1. Judges have no strong basis for gainsaying the complex business decisions made by a BoD for a particular business in an exacting set of circumstances.

2. Judges realize that it would be unfair to become Monday-morning quarterbacks. Hindsight is very easy and largely worthless and the law knows this.

3. The power of precedent, so strong in many other legal areas, is difficult to apply precisely in the highly complex and ever-changing world of business decision-making.

4. As a practical matter, large companies with thousands of shareholders will always have some shareholders who are (or who claim they are) unhappy with some decision or other. The law cannot permit them all to sue based only on a disagreement, after the fact, with a business decision made by a BoD.

To cite a counterexample, if it can be shown that a particular Director (or the BoD as a whole) knowingly led employees to believe that they were going to be paid for their continuing labors while at the same time signing papers leading to the dissolution of the corporation, that could be (and should be) costly in a personal way to that Director (or to that Board of Directors).

Employee pension plans furnish an instance of where there is an especially complex financial commitment—obviously lasting many decades. Small companies or startups (Chapter 12) should give pension plans a wide berth and should substitute some other compensatory mechanism for the missing company pension contributions.

13.3 THE GOVERNANCE TEAM

If the previous discussion has turned off some potential engineers as aspiring Directors—if some readers have made a mental note to not, ever, join the BoD of any company, even one of interest—that has certainly not been the intent. An analogous situation might be to join a mountain climbing team or a soaring club. As with these latter endeavors, there are risks; it would be foolhardy not to learn about them and take steps to mitigate them. Rushing in to such activities in a naive way would be foolish, but many still wish to enjoy them.

Guarding against such a pollyanna attitude towards joining a BoD was largely the point of the last section (§13.2). This section now provides balance by showing, broadly speaking, what BoDs do, and consequently why it is an exciting place to be. Only the most interesting and important corporate issues are discussed at the BoD level, and it is there that one can integrate all one's managerial and technical[21] experience into a coherent, stimulating senior management experience.

Critical BoD Roles

The duties and responsibilities of a corporation's BoD are largely specified by law, with some interpretation left to best practices, and with the final word given by the courts. There is, as usual, some variation in how many of these duties are critical, but the short list of Leighton and Thain,[22] shown in the accompanying box, is as good a place as any to start:

Responsibilities of Corporate Boards of Directors

1. Overseeing Strategic Management
2. Selection and Evaluation of Senior Management (especially the CEO)
3. Shareholder Relations (only the BoD can declare dividends)
4. Protecting and Exploiting Company Assets
5. Fiduciary and Legal Requirements

---

[21] Assuming, as always, that most readers are engineers.
[22] DSR Leighton, DH Thain, *Making Boards Work*, McGraw-Hill Ryerson, 1997.

We have already introduced Responsibility #5 in §13.2, and will be discussing Responsibility #1 presently, in §13.4. Hence we shall spend our time here (§13.3) on Responsibilities #2, #3 and #4. The complete list of responsibilities is certainly longer and further nuances on what these "responsibilities" mean in practice are given in the references. The principal point being made in this book is this: Directors of corporations are all people who have, based on their inherent ability and their pro-active career decisions, ended up on a Board of Directors. While these positions are[23] not, frankly, for youngsters (since they emphasize experience and wisdom), a wide spectrum of professional backgrounds is desirable to elicit as many fresh ideas as possible and to avoid the paralysis of groupthink.

BoD Makeup

How does one choose a BoD? Several comments have already been made on this subject—including the arguments (made above) that there should be more engineers as Directors—and the general premise that a broad philosophical and professional representation (provided solid business experience is proven) is preferable to a narrow viewpoint. Very few Directors have had explicit training on how to *be* a Director, although some business schools offer such training now in their executive course programs.

One subject endlessly discussed by BoD theorists is the subject of *inside* vs. *outside* Directors.[24] The point is that the Outsiders are more independent and are more likely to represent all the shareholders; management reports to them, not the other way around. On the other hand, Insiders know much more about the details of the company and can thus provide more complete and timely in-

---

[23]There are a few exceptions to the Youngster Rule, among them being this: The youth of the Company Founder should not be an issue.

[24]There are several (related) definitions of Inside and Outside, depending on the writer and the jurisdiction. Some say that an *inside* Director is one who is also an Officer of the corporation (while an *outsider* is not); some say that an Outsider draws no compensation from the corporation other than his Director fees and expenses (while Insiders do); and some say that an Insider is one who owns more than X% of the company (while Outsiders do not). We shall not fuss about these definitions here. The slight variations do not obscure the main point about the need for independence.

formation on which to base important decisions. In practice, both Insiders and Outsiders are needed.

As the company size grows and the BoD size grows with it, the ratio of Outsiders to Insiders will grow. It is rare that more than two or three Insiders[25] are needed. It would frankly be absurd were the CEO not on the BoD to provide the vital link both up and down (see Fig 13.4). A CFO is often also on the BoD, as is the COO, if this position exists.[26]

Another obvious question is, "How large should a BoD be?" Smaller companies tend to have smaller BoDs (they cannot afford a larger one), but for mid-size to larger companies, Salmon advises[27] that "between 8 and 15 members is probably about right. Fewer than 8 Directors cannot staff [BoD] committees with enough outside directors. But more than 15 members almost always [is unwieldy]."

Women On The BoD

Most BoDs are populated primarily by men. Women's rights groups have sometimes spoken of a "glass ceiling," referring to a level of management and leadership responsibility above which women are not permitted to rise. Not precisely defined in these claims are either the exact nature of the inhibitory mechanisms or a specification of precisely who is doing the forbidding and why. However, the basic statistics are clear enough: The dominant gender group on BoDs is overwhelmingly men.

This issue is highly charged politically, as well it should be. However, the methods of the appertaining social sciences shed more light on this phenomenon than the heat of political rhetoric. Some women say that a woman has to be twice (thrice? more?) as

---

[25] If advice or information from other senior managers is needed, they can al-ways be asked to give a presentation to the BoD.
[26] It would be admirable if a CICO (Chief Intellectual Capital Officer) were a common position in modern knowledge-intensive companies, and on the BoD. The ideal scenario would be if the CFO and the CICO, in a balanced strategy and in concert with the CEO, could develop a strategy for making the process depicted in Fig 10.1 work.
[27] WJ Salmon, "Crisis Prevention: How to Gear Up Your Board," *Harvard Business Review*, Jan–Feb 1993.

good as any man to "get ahead" at these management levels. Some men say that women who become senior managers or Directors were so appointed at least partly from a desire for gender equality. While it is impossible for these two positions both to be correct at the same time, it is quite possible for them both to be wrong.

Since Directors are normally chosen from among individuals who have had a long and distinguished record of senior management, it seems likely that, as more and more women enter and excel in professional fields leading to senior management—including engineering—there will be, after a transitional period of time, many more women who serve as corporate Directors. It will take a few generations for the choices and cultural norms of girls early in elementary school to change their course emphases and self-image and to diffuse through the many levels of the system.[28] Fortunately, modern demographics suggest that many women are now engaged in the kinds of advanced education, including engineering, that lead to these senior managerial opportunities[29] and they are currently doing very well indeed. Perhaps some day we may become concerned about why there are so few *men* on BoDs!

There may be differences here between the viewpoints of younger vs. older women, and possibly also among women in various professions. The only subset with which the author has had extensive mentoring experience is that of the many super-charged young men and women in modern engineering schools. The attitude of these young women (which is somewhat at odds with the standard feminist stance) is often as follows: "Please don't give me any special breaks. I don't need them and I don't want them. They will haunt me throughout my entire career. The feminist canard that I have to be twice as good as a man to be considered equal may have been true sometime, some-

---

[28] The CCPE (Canadian Council of Professional Engineers) has data that indicate that the enrollment of women in engineering, after slowly growing for many years, may be leveling off. Since the number enrolling in other, better-paying professions is still growing, the inference is obvious.
[29] Evidence that this is already happening is provided by the Conference Board. From 2000 to 2005, the number of manufacturing companies with one or more women on the BoD went from 64% to 73%. (Corresponding figures for the service sector are 67% to 80%. The financial sector went the other way, going from 75% to 69%.)

where, for somebody, but I have no evidence that it is true *now*, *here*, for *me*. Just, please, let me compete as a human being, without the needless gender complications. I know I will do well in my career."

It will probably take several generations, as it does with other natural processes, for the now-lively interest of women in the subjects of management and engineering gradually to populate the many higher levels of corporate leadership, thus creating a rich pyramid of human talent from which the best women, and the best men, can percolate up to the highest levels in corporate governance.[30]

In the critical matter of assessing who is best, however, it would be advisable to broaden one's perspectives somewhat. There is an increasing awareness that men and women are not interchangeable[31] (surprise?) and that, although the leadership styles of women and men appear not to be statistically identical, neither style is superior to the other.

Both styles of leadership, to the relatively minor (but personally and politically important) degree that they are distinct, are much needed when they are inspiringly executed. And, by a providential coincidence, the special genius that many women have for relationship building (as compared, perhaps, to the comparable genius than many men have for competitive goal achievement) is precisely the new basket of ingredients that knowledge-based, new-economy companies need to nourish their sources of intellectual capital (people).

Subverting Outside Directors

Until recently, many BoDs did not do their job very well, often with tragic consequences for the shareholders. They clearly had the legal

---

[30] Just after this chapter was written, the leftist government of Norway announced that all companies in that country must have at least 40% women on their boards by 2008 or be closed. That's one way to do it! (There was no mention of a minimum for men.)

[31] Early and radical versions of feminism tried to claim that men and women were identical and that quotas should be established before the end of the week to mandate equality in any profession where the number of men exceeded the number of women. More recent discussions, thankfully more dispassionate and evidence-based, make the important point that the key issue is the quality of, not legally-mandated quotas for, leadership. See, for example, S Helgesen, *The Female Advantage: Women's Ways of Leadership*, Currency Doubleday, 1990.

authority to exercise the needed power, but often neglected to do so. One reason for this state of affairs was that many of the Outside directors were more inside than they appeared. Consider the following scenario:

> The BoD of Company A (Co-A) has an opening for adding a new Director. The CEO of Co-A knows many senior managers in other companies and he chooses to ask the CEO of Co-B to join his BoD. On paper, CEO-B is ideal for the task. After all, who could better help run a company than someone who is already running a company? CEO-B is also an "outside" director—no salary from Co-A, no (or negligible) shares in Co-A; and not an Officer of Co-A. So CEO-B joins BoD-A.
>
> Not long after this appointment, Co-B also needs to replace a retiring Director, and so CEO-B asks CEO-A to join BoD-B. Again, a highly competent outside Director is added. What's wrong with this picture?
>
> What's wrong is that the decision of who should join BoD-A should be decided by BoD-A, not merely by CEO-A. And similarly with BoD-B. When CEO-A asked CEO-B to join "his" BoD, he likely did not mean "his" in the sense of "the BoD to whom I report." He probably meant ownership, as in "the BoD that's in my hip pocket."
>
> As the number of these cross-appointments grows, and as more and more "outside" Directors feel that they are part of an exciting network of Directors whose mutual back-scratching is the valuable currency (since they all owe their appointments mutually to each other), the chances grow increasingly dim for truly independent advice and due diligence. The fiduciary responsibility to the shareholders can slowly become a non-fiduciary deference to the CEO.

To avoid this *schlamassel*, many large, sophisticated BoDs have a *nominating committee*, whose sole purpose is to select new Directors for the BoD. A second relevant prophylactic against the game of musical chairs for CEOs is the question of whether the CEO and the Chairman of the Board are two persons or one, a subject to which we now turn.

Chapter 13: Governance

Chairman & CEO: One Role Or Two?

In brief: The role of the CEO is to run the company; the role of the Chairman[32] is to run the Board. There have been many instances where a one-person, dual-role Chairman and CEO has received an all-thumbs-up for his performance in leading his company. Lee Iacocca at Chrysler and Jack Welch at General Electric come to mind, and there are many others. However, despite these success stories, most students of governance believe that, just as an enlightened despot may be tolerable or even welcome in some situations, it is usually difficult to assess until much later whether despotism or enlightenment was the more prominent feature.

One way to arrive at the two-role model is to list the duties of the Chairman in leading the BoD. These include[33] setting BoD meeting agendas, managing BoD meetings, structuring BoD committees, working with these committees, working one-on-one with the CEO and other Directors, managing BoD performance, and developing the BoD (e.g., the committee structure and attracting new Directors). For all but the smaller sized companies, this sounds like a full-time job, not one the CEO does in his spare time.

Another consideration is that the Chairman provides a buffer between the BoD and the CEO. It is the Chairman's job, for example, to set the meeting agendas. While this should be done after seeking advice from the CEO (and other Directors), the final decision rests with the Chairman. If the CEO wants to suppress discussion of a certain important subject, that will be easily accomplished if he is also Chairman, but more difficult to pull off when the Chairman is a different person. As another example, the Chairman should be an *ex of-*

---

[32]Some feel that it is politically incorrect to use the generic Chairman as a gender-neutral noun, and indeed Chair has largely replaced Chairman in the public and not-for-profit sectors of society. We shall not follow this precept, however, because throughout business—which is what we are talking about—Chairman continues to be used as a gender-neutral title for this position. For example, Carleton (Carly) Fiorina, Chairman and CEO of the high-tech company Hewlett Packard until February 2005, was referred to (by others and by herself) as Chairman. (As an aside, she was not an engineer; her first degree was in medieval history and philosophy, from Stanford.) The author does not feel comfortable arrogating to himself the authority to re-title the top management positions in all the *Fortune 500* companies.
[33]According to Leighton and Thane, *op. cit.*

ficio member of all BoD committees to ensure coherence between their information, assumptions and activities. If the Chairman is also the CEO, however, he can't (or certainly shouldn't) sit on the *compensation committee*, one of whose primary tasks is to decide his compensation!

Another example, and a very practical one, centers on BoD meetings themselves, at which more significant information is provided by the CEO than from any other single source and during which the CEO does more of the talking than any other Director. Yet it is well known (and common sense) that no one can chair themselves. While the CEO is focusing on presenting key information and answering difficult questions, the Chairman can keep an eye on the general flow of the meeting. He can ensure that other viewpoints are exhibited, that the agenda's priorities (which may be slightly different from the CEO's priorities) are the basis for time allocation, and that the meeting accomplishes its overall objectives.

Still, Lorsch[34] cites the statistic that the chief executive is also the Chairman in more that 80% of [U.S.] publicly held corporations. The incidence of this double-agent setup is much less in Canada (the two roles are separated in two-thirds of companies) and Europe. It goes without saying that, if a two-person tandem is selected, good chemistry is required between the Chairman and the CEO to move forward with positive energy.

BoD Committees

Two committees of the BoD have already been briefly mentioned—the nominating committee, and the compensation committee. The most important single committee, however, is the *audit committee*, whose invention is surprisingly recent but whose importance grows daily with each accounting scandal.[35] When Enronitis breaks out somewhere, there is plenty of blame to go around; the usual suspects include the CFO, the CEO, the Chairman, the outside fi-

---

[34] JW Lorsch, "Empowering the Board," *Harvard Business Review*, Jan–Feb 1995.
[35] In fact, recent U.S. legislation is starting to give the audit committee, not the full board, final say on some related matters, such as its audit policy and its committee budget. The audit committee is assumed to be dominated by Outsiders.

nancial advisors, the outside auditors (too frequently being these selfsame outside financial advisors!), the full BoD and the audit committee. The importance of the audit committee is growing in prominence as its role continues to evolve.

Several other committees are also common, including a governance committee, an executive committee (for very large companies), and so on. More detailed specifications of their roles will be left to the references.[36]

When committees report to the full BoD, as they should do with some regularity (e.g., every quarter), a delicate balance must again be struck. If the BoD insists on rehashing everything done by each committee, or if some individual Directors feel that they must start to examine every committee decision in complete detail *ab initio*, the committee system will have collapsed; it is supposed to improve efficiency and effectiveness, not create more work that is subject to multiple layers of scrutiny. This principle, which applies not only at the highest management level—governance—but at all lower management levels as well (except, perhaps, to apprentices, who are not assumed competent) is shown in Fig 13.6, whose theme is of kindred spirit to Fig 11.4. It shows four categories that can be used to describe *a particular decision*.

Figure 13.6: Broad Categories for Reviewing Decisions of Others.

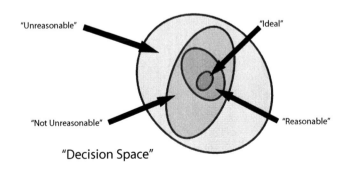

---

[36]B Lechem, *Chairman of the Board; A Practical Guide*, Wiley, 2002.

At the center is a region called "ideal." Note first that it is a region, not a point, because there is no absolutely perfect decision, and there is no absolutely perfect person to bestow that label. There is only a small region into which would fall the decisions of the very best people (at making this particular kind of decision). For each individual reviewing the decision (each Director, in the present instance), there is a point, but for the BoD as a whole, there is a region, and no Director should consider himself perfect.

Next out from the center is a region called "reasonable." In a good organization, this is where most decisions fall. Yes, with further financial, human and temporal resources, the decision could have been refined further, so as to be moved into the "ideal" region, but all those resources cannot be thus expended unless the decision is a critically important one. Directors have to know what is most important and what is less so.[37]

The next region, described as "not unreasonable," sees the decision as flawed, but not fatally so, unless the decision is of the highest import. Some comments will likely be made about how the decision[38] could have been improved, but the descriptor "not unreasonable" means—and this is the main point of Fig 13.6—that the decision is not normally revoked or changed. While we should normally hope for a "reasonable" decision since "ideal" ones are quite rare, the system simply doesn't work if all decisions that are "not unreasonable" are re-worked and changed. Still, if an individual becomes known for always making decisions that, while "not unreasonable," are rarely better than that, that person should not expect to be promoted to a position where decisions are even more

---

[37]This remark most emphatically does *not* imply a relief from the process of *due diligence* (see §13.2). Indeed, the process described here for deciding on the reasonableness or appropriateness of corporate decisions is not antithetical, but is in fact identical to due diligence.

[38]Perhaps an analogy to the court system may be helpful here. Rarely, if ever, does a juror know with *absolute certainty* whether the defendant is guilty or innocent. So one is never at the exact center of the decision space in Fig 13.6. A juror must decide whether he or she is very, very close to the center (a balance of probabilities described as being certain *beyond reasonable doubt*) or farther away, in which case the juror in criminal law must not find the defendant guilty. Newspapers sometimes claim that a defendant was found "innocent" but that is a regrettable and confusing fallacy. Being found "not guilty" in the court system is certainly **not** being found innocent!

difficult and even more important.

Finally, we have the decisions[39] that are found "unreasonable," meaning that the BoD will either change them or ask that they be reworked. BoD committees that make such decisions more than rarely should have their processes or membership adjusted. Managers and Officers whose decisions are thus regarded, especially if a pattern becomes discernable, may be vulnerable to career realignment.

CEO Evaluation And BoD Evaluation

To continue with the subject of evaluation, there is no higher duty for the BoD than the selection and evaluation of the CEO. Evaluation is a somewhat sensitive area, since the CEO is one of the Directors (if not, why not?) and few CEOs welcome this potentially fractious process. Yet such evaluations are essential because they have a beneficial impact[40] and because they are a basic BoD duty. At the very least, the Chairman should arrange[41] that an annual questionnaire be put to the Directors with questions concerning the leadership, operational effectiveness and financial performance of the CEO.

The CEO should also respond to the evaluation questionnaire, although this response should be handled in a slightly different fashion. One process that works is this:

(a) No Director, especially the CEO, is aware of the responses of any other Director.
(b) The Chairman synthesizes all responses (including his own and the CEO's) and presents his written report to an *in camera* phase of the BoD meeting.

---

[39] In reality, and unlike Fig 13.6, the degree of unreasonableness of which decision-makers are capable may be unbounded.
[40] JA Conger JA, D Finegold, EE Lawler (III), "Appraising Boardroom Performance," *Harvard Business Review*, Jan–Feb 1998.
[41] CEO performance evaluation is yet another instance where having two individuals serve in the two roles of Chairman and CEO makes the cheese less binding. Lorsch (*op. cit.*) says flatly that CEO evaluation "is a major step toward empowering the board because it delivers a clear message to both the CEO and the Directors that the former is accountable to the latter."

(c) During the *in camera* phase, an organic (consensus) BoD response is constructed, ideally without consuming great gobs of time and without recalcitrant minority opinions or outlier interpretations.

(d) The CEO rejoins the meeting and the Chairman summarizes the consensus evaluation, stressing pieces that were done well and areas that need improvement. He then invites additional comments from the other non-CEO Directors, elaborating on this consensus evaluation.

(e) The CEO is then invited to add his own views on his performance, *likely moderated by the consensus evaluation*. There may be significant achievements of which the other Directors are not aware, or noteworthy excuses for troublesome disappointments.

(f) Finally, some high-level plan for additional CEO performance improvement is identified. The Chairman should prepare a written version of this plan and present it to the CEO as expeditiously as possible.

The author is of the opinion that no BoD activity reveals more perceptively the health of the governance process than CEO evaluation. If the BoD cannot generally agree on the CEO's performance, or if the CEO cannot concur with the BoD on what has or has not gone well, then the quality of governance is in jeopardy.

To move to the subject of evaluation of the BoD itself, effort on this front[42] should also be made at least annually. There are several meanings to "BoD evaluation," depending on precisely what is being evaluated and by whom. Meant here is a process through which all Directors are asked for their opinion on how they view the

---

[42]Read, for example, G Donaldson, "A New Tool for Boards: The Strategic Audit," *Harvard Business Review*, Jul–Aug 1995.

health of BoD processes and the quality of BoD outputs. Through some combination of written questionnaires and private conversations, the Chairman should elicit the opinions[43] of all Directors[44] on how things are going. These evaluative remarks would then be considered at an upcoming BoD meeting, thus providing helpful mid-course corrections to the BoD trajectory.

Some readers might demur from this BoD-evaluation process on the grounds that it is essentially a self-evaluation, which poses dangers of subjectivity. This is a reasonable and serious concern. If the BoD is a Country Club Board, in the language[45] of Fig 13.7, one should not expect much from the BoD self-evaluation process. (One should not even expect that the process exists within this mentality.) Fortunately, human beings tend to be more objective about their team characteristics than they are about their personal characteristics. A high-voltage team of professionals—which is what a BoD is supposed to be—can usually refine and improve their performance through self-criticism; they should at least try to do so at reasonable intervals.

Figure 13.7: The Emotional Climate of Boards.

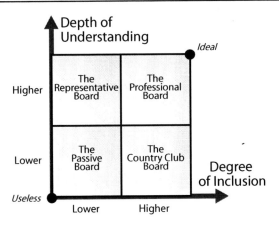

---

[43]Salmon (*op. cit.*) provides "22 Questions for Diagnosing Your Board," a framework for BoD evaluation.
[44]One presumes here that all Directors are professional to the extent that they are familiar with the legal responsibilities of both the BoD as a whole and of individual Directors.
[45]B Garratt, *Thin on Top: Why Corporate Governance Matters and How to Measure and Improve Board Performance*, Brealey, 2003.

Shareholders vis-à-vis Stakeholders

The picture painted by Fig 13.5 (and used in all the discussion above) is that the BoD and the auditors (elected by the shareholders) are responsible solely to the shareholders. While this remains substantially the case, others think that the "best interests of the corporation" means responsiveness not just to shareholders (a.k.a. stockholders, or owners), but to a wider group, called *stakeholders* (Fig 13.8).

Some of these stakeholders are well-documented and protected[46] in the financial statements (e.g., banks, other lenders, suppliers) or by contract (e.g., customers, insurers). Employees are also protected in terms of Balance Sheet liabilities and through extensive employment legislation in most jurisdictions. Auditors and Directors are usually well paid for their services and should expect nothing more.

Figure 13.8: Other "Stakeholders."

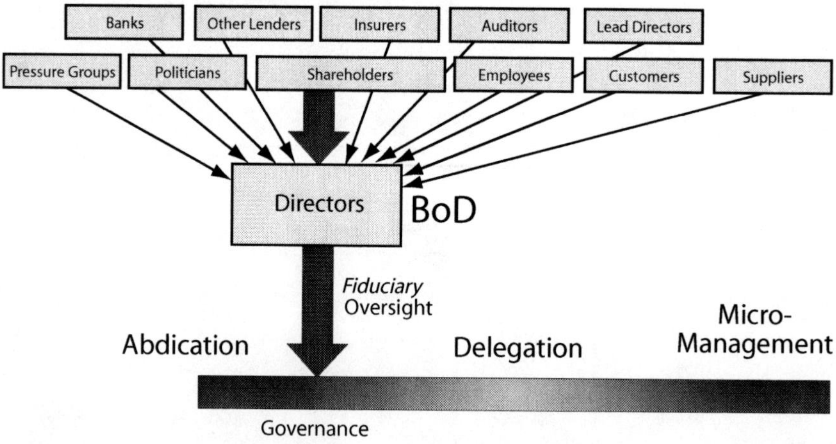

That leaves politicians (i.e., the government) and pressure groups. Certainly, anywhere there is a concentration of wealth, there will be those who will try to access it; fortunately, most of

---

[46]One category not explicitly mentioned in Fig 13.8 is relatives or relations [of people with real power on the company]. Family companies are usually lifestyle companies and are not treated herein, except perhaps obliquely in Chapter 12 on startups. It is not clear why anyone would want to work for a company where genetic relationships trump objective job performance. Nepotism is one of the killer viruses of any meritocracy, including corporations.

those folks[47] will use legal means, including the ballot box. The formal way for people who have no official connection to a corporation to acquire the benefits of its earnings is through the tax system, and there is a certain fairness to this arrangement—at least up to a point (and the argument always centers on precisely where this point lies)—inasmuch as companies benefit every day from numerous components of infrastructure to which they would not have otherwise contributed. (In addition to direct tax on the corporation, its employees and suppliers also pay tax.)

On one point there can be no rational debate: much as it may give dyspepsia to Michael Moore types, corporations are legal *persons*, and they must obey the law. Further, just as with human persons, they must strive to do more: they must be *ethical*. Figure 13.9 (which bears an intentional resemblance to Fig 13.6) shows the relationships between legality and ethics. Illegality may have some transitory benefits, quickly squelched; unethical behavior may offer some short-lived gains, with no lasting rewards; the only ethical strategy than can build great companies is zero tolerance, everywhere, for shady behavior. The proceedings in and from the Boardroom have a profound influence on the ethical stance of a corporation.

Figure 13.9: Broad Ethical/Legal Categories.

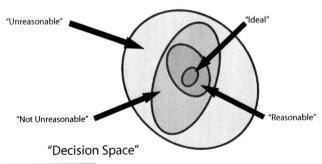

---

[47]As in all complex public issues, some special interest groups go well beyond economical rationality in their demands on corporations. For example, the claims of some seem perilously close to contending that companies have some obligation to hire (or retain) employees they don't want or need. The irony is that most of these claimants have parents whose financial welfare depends on the continued profitability of the very corporations they are trying to force into bankruptcy.

Further Remarks On Governance

A great deal of additional information on corporate governance is available from the sources cited above. In fact, if one peruses all the references given in this chapter, one will have taken a strong course on the BoD and its Directors. As with any team activity, the best teams are not just those boasting players with high levels of individual talent; they also have great *chemistry*, a subject concerning which no general book can make predictions in any particular case. Still, if the Directors share common goals and values, this will raise the likelihood of catalytic chemistry.

One danger that should explicitly be mentioned in closing is the over-emphasis on *process* over *results*. With all the discussion of legal liability in §13.2 and all the examination of process in this section (§13.3), there may be the impression that the BoD is a bureaucracy-in-miniature, paralyzed and hidebound by the endless processes prescribed for each meeting. On the contrary, a substantial portion of BoD discussion should be extemporaneous, topical, and responsive to immediate issues (strengths, weaknesses, opportunities and threats). Here is a concrete example:

> Just above, we had a look at both CEO evaluation and BoD evaluation, both of which were touted as being highly important. The next level in this process might be the evaluation by the BoD of individual Directors. While this is not a bad idea *per se*, it was not introduced above as an important part of the governance process simply because to do so would (in the author's opinion) imply that the line would be crossed from being results-based to being process-based.
>
> There are many other ways of dealing with suboptimal Directors, including outvoting, term appointments (three years is conventional), and other more subtle methodologies. Every hour a BoD spends introspectively navel-gazing is an hour that they are not spending on their solemn duties to govern the corporation.

Lastly, it has been repeatedly pointed out in this book that no two businesses are the same. This is what makes business success simultaneously so frustrating and so intellectually challenging. One size does not fit all, and the many distinct parameters of each company will lead, naturally and properly, to divers CEO and BoD styles.

13.4 THE STRATEGIC IMPERATIVE

Of all the major responsibilities and goals placed on Directors as part of their governance in a BoD, the only one that has not been explicitly alluded to above is the responsibility for the strategic direction of the corporation. We close this chapter with some comments on strategy, including a brief dissertation on a strategic element of particular interest to engineers—technology roadmapping.

The Endless Quest For A Good Strategy

While trying not to get bogged down in semantics, it is still advisable, as always, to say a word or two about what is meant by *strategy*. A fun example is the game of chess, wherein the immediate questions concerning who takes which pawns or pieces in the next move (or two or three) represent the *tactics* of the game, whereas longer-term issues[48] comprise the strategy. Beginners always find it takes hundreds of games to begin mastery of the tactics, much less thinking about strategy, a glimmer of which takes many more hundreds of games. A more pedestrian example of strategy vis-à-vis tactics will be given below, in the discussion of technology roadmapping.

---

[48]Such strategic issues would include, for example, whether to create a closed game (everything blocked by everything else) or an *open* game (a firefight over long files, ranks and diagonals); whether to initiate a pawn-storm down the right or left side of the board; and whether to castle on the king-side or the queen-side. The distance from the chessboard to the board room is not all that far, philosophically. A further analogy between business strategy and the roughly three phases of a chess game (the *opening*, the *middle game* and the *end game*) is slightly more nuanced, since the business is not supposed to end, but the metaphor for particular strategic business *initiatives* is intact. Also noteworthy is that there are huge chess tomes on both the opening and the end-game phases, but the discussion of the middle game is largely left to commentary on the matches of former world champions, some of whom were known as tactical geniuses and some as strategic geniuses (although all were in the tiniest top percentile in both). It seems that the middle game, whether in chess or in business, is inexplicable and ineffable: Only through watching geniuses at work can one hope to glean the most subtle insights. In business school, this is called the "case method."

An enormous body of written opinion exists on the subject of business strategy. While this circumstance is undoubtedly due in part to the intrinsic importance of the subject, it may also be traced to the inherent difficulty of saying something intelligent yet broadly applicable about all businesses (given their perplexing range of contexts) and to the intrinsic impossibility of prophesying the future without looking silly (for some of the attendant risks, see Chapter 7). What is more, strategizing is exactly the sort of fundamental, abstract, high-value task that poses an endless intellectual challenge to the academic component of the business community—and this drives the creation of an unusually rich body of literature.

The best recent overview of the philosophy of business strategy is due to Mintzberg[49] *et al.*, who have identified no less than ten "schools" of strategic management, some with recognizable variants. These ten contenders vary greatly in their age and total literary blossoming. Some schools are more concerned with the strategizing process; others are focused on results and content. The important point in our own discussion here is that the BoD is ultimately responsible for the formation and approval of corporate strategy.

Vision, Mission, Purpose . . .
To give but one oft-used example of process, some have suggested that one should start by identifying the corporate Vision, Mission and Purpose (or similar words), and from these grand pronouncements everything (it is argued) should logically flow. Engineers naturally fall in love with this approach, since it suggests that there are some profound statements that, once found or once promulgated, contain all the nuggets of truth one ever needs to know—rather like knowing the Lagrangian function for a system of 1,000 connected bodies, from which, when this one function is known, one can derive all the governing equations of motion.

---

[49] H Mintzberg, B Ahlstrand, J Lampel, *Strategy Safari: A Guided Tour through the Wilds of Strategic Management*, The Free Press, 1998.

The author's opinion, having embarked on several such philosophical voyages, is that this process is, at best, tenuously related to what needs to be decided and, at worst, fatuous. Whole countries have difficulty with their vision and mission, so perhaps companies should get on with less sweeping issues, like **(a)** their survival, and **(b)** what their growth strategy should be. To say that one's mission is to make a pile of money, or be the best darned company on the east coast, or to blaze new trails in nanotechnology (or whatever) is, by itself, merely grandstanding. The resulting logical flow isn't very logical and doesn't flow. As in so many other instances, actions speak louder than words.

SWOT (Again)

At the good end of the common sense spectrum, we have SWOT[50] analysis—first introduced in §7.6 and Fig 7.10—which simply requires that one spend some time identifying what is good and bad, both inside and outside the company, and making rational inferences therefrom. It is difficult to imagine any potent strategic process that does not include some version of this SWOT process, perhaps with special flourishes and appendations, but still grounded on these four basic baskets of facts.

The BoD and its individual Directors must be aware, at least in broad brushstrokes, of the primary battle readiness characteristics of the corporation (both good and bad), and through further due diligence they must seek information also on the relevant environmental factors that are conducive (or not) to business.

As shown in Fig 13.10, the strengths and weaknesses exercise should at the very least produce a disclosure of the corporation's core competencies—those value-added activities that the company can perform (especially in combination) better than its competitors. If it proves to be difficult to identify such competencies, the exer-

---

[50]For readers who have recently joined the discussion, the acronym SWOT connotes "Strengths and Weaknesses" (an inward-looking exercise) and "Opportunities and Threats" (an assessment of the external business setting, including the competitive environment).

cise has already shown itself exceedingly useful: the company has been revealed as incompetent. (This is not an orphan disease for any business. What is needed in this case is not just a strategy, but a more fundamental re-tooling.)

Figure 13.10: Strategic Plan (SWOT is Just the First Step!).

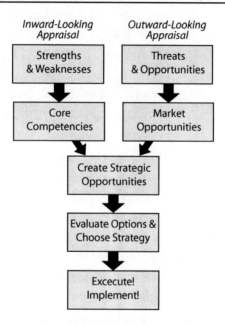

This exercise is the first half of a rubber-hits-the-road valuation of the company's intellectual capital (Chapter 10). The second half is the external assessment (threats and opportunities), which should lead to some candidate market opportunities. Once again, if no serious market opportunities can be identified, this presages serious trouble—although it must be said that, unlike the internal assessment, which should be relatively simple and accurate, the external assessment is almost infinite in scope and less precise in its conclusions. A negative result in the search for business opportunities may simply indicate that more thinking is needed outside the infamous[51] box.

---

[51] There should be a law requiring anyone who recommends "out-of-the-box thinking" to pay a stiff fine if he or she does not also carefully define "the box."

The core competencies exposed by the internal examination of strengths and weaknesses should be matched against the market opportunities discovered in the external research on opportunities and threats. Once again, if nothing interesting can be discerned, the SWOT process has done its job: Stop rowing toward the cataract; abandon the current course; that way lies suicide.

More optimistically, we assume that *several* potential business strategies are suggested (Fig 13.10) and that a selective process of evaluation-and-choice is required to make the final strategic decision. As always, we must remember two fundamental things about a strategic plan, whether the context is sports, chess or business: (a) the Plan will become obsolete shortly after[52] its formulation; and (b) the Plan is only the first 5% of the enterprise, the other 95% being successful implementation (execution).

BoD Role In Corporate Strategy

We continue the discussion with the realization that the BoD has a mandate for setting corporate strategy. This does not mean that senior management should sit back passively, waiting until the next BoD meeting to learn what strategy will be passed down from on high. To the contrary, it is the primary responsibility of senior management not only to execute the current strategy, but also to develop future strategic alternatives. The role of the BoD is, as they say in the U.S. Senate, "to advise and consent." The CEO and his management team cannot leave strategy development either to a more senior management level (governance), or to some less senior group[53] of strategic planners.

---

[52]Some senior managers may be tempted to say this: "Since the Strategic Plan cannot be relied upon for five years [or whatever the time horizon is intended to be], why bother spending time preparing it?" These folks have missed the whole point and are the sort who are perpetually surprised at almost everything that happens. This makes their life quite interesting and the life of their company stakeholders much *too* interesting.

[53]Two or three decades ago, when the pace of business change was less frenetic, some managements left the matter of strategic planning to an internal side-committee to develop, and some—difficult as this may be to believe in the 21st century—*contracted out* this critical and protean activity, seen back then as something of a frill. This dysfunctional view of the aims and processes of strategic planning usually had the chilling outcome of placing a very large report—and the company that commissioned it—up on a dusty shelf.

Specific strategic advice is, by definition, not possible here for particular companies. However, it should by now be clear that, in the words[54] of a key reference,

(a) Boards should be constructively engaged with management to ensure the appropriate development, execution and modification of the company's strategy.

(b) The extent and nature of the board's involvement in strategy will depend on the particular circumstances of the company and the industry in which it is operating.

(c) While the board can—and in some cases should—use a committee of the board or an advisory board to analyze specific aspects of a proposed strategy, the full board should be engaged in the evolution of the strategy.

Clearly, strategy identification is a top BoD priority; the BoD cannot just sit back and blame others (a potentially long list) if the company strategy (assuming it has one) fails to generate the desired business results.

Technology Roadmapping

To conclude this section, this chapter, and this book, we briefly describe a process that is intrinsically strategic, and one in which a Director with technical training (and due diligence) should be an *invaluable* asset on the Board of Directors of any company that plans to use new technologies either in its internal business processes or as value-added components in its product offerings to its customers.

If every BoD should have[55] at least one outside Director with strong business experience (self-evidently a requirement), and at

---

[54]See *The Role of the Board in Corporate Strategy*, Report of the NACD (National Association of Corporate Directors) Blue Ribbon Commission, NACD, 2000.
[55]Sometimes more than one role can be played by the same person.

Chapter 13: Governance

least one outside Director with accounting experience (a candidate to chair the Audit Committee), and at least one outside Director with legal experience (someone who can assess risk exposure, can peruse key contracts, and who is a game player in the best and most complex meaning of the word), most modern companies should also have a Director who is technologically literate. An excellent example would be the process of *technology roadmapping*, the part of the strategic planning that specifies what technologies the company should exploit, develop, or license-in over the next (say) five years.

These roadmaps are constructed by identifying all the existing and upcoming technologies, and then extrapolating to find what constellation of technologies the company must have five years hence. The ones that are both needed and missing must either be developed in-house, acquired from outside, or contracted out.

There can be several levels of roadmaps, including sector roadmaps at the national level. For example, there are national roadmaps for information[56] technology (IT) and for several other key platform technologies as well. Frequently the learned societies in the cognizant disciples assist in these programs. Motorola, Mitsubishi and Ford were early developers of this important facet of strategic planning. There is also some[57] roadmap horse-trading between large high-tech companies, so that both sides can improve their maps.

An engineer on the BoD of a corporation would not be expected to personally produce a detailed technology roadmap for the company, but would prove invaluable in aiding other Directors to oversee and diligently examine the roadmap(s) prepared under management supervision.

As time goes on, leading corporations will begin to realize more poignantly that it is not sufficient just to get their accounting right,

---

[56]The best-known example of a successful long-term technology growth extrapolation is Moore's Law (not a "Law" in the same sense as the Law of Gravity!) which states that the growth of computing power per dollar is exponential, with a doubling time of about two years. This law deserves to be well known because it has held approximately true for over three decades.
[57]See, for example, P Groenveld, "Roadmapping Integrates Business and Technology," *Research and Technology Management*, Sep–Oct 1997, pp 48–55.

nor just to optimize their capital structure, nor just to have effective human resources policies, etc. They will also see that their sustained competitive advantage and their ability to generate profitable operations well into the future—indeed, their ability to survive and thrive in an environment of revolutionary technology development—depend also on their skill at superior technology forecasting and on integrating these forecasts seamlessly into their overall business strategy.

As a final observation, the current globalization process and the consequent outsourcing of jobs to developing countries with lower wage rates can only be successfully countered by obtaining higher productivity from the well-paid labor in the developed countries. Significant improvements in productivity always rely on technology advances to be realized.

In §5.4, we defined technology as "any man-made thing that one can use to assist one in doing something one wishes to do." Now the deep and relevant truth of that definition is obvious. New technologies, wisely chosen and economically developed or acquired, are the keys not only to being able to market higher-tech products, but also to ever-higher productivity in operations. The miracles mentioned (tongue in cheek) in Fig 3.12 turn out not to be miracles after all, but are instead the consequences of a strong intellectual-capital balance sheet: good people, who are well managed, and equipped with the latest technology. Engineers should have vital roles to play in every facet of such companies: in design; in research and development; in management at all levels; and on the Board of Directors.